Popular Mechanics

GUIDE TO MODERN TELEPHONES

Hearst Direct Books
New York

Acknowledgments

This book is published with the consent and cooperation of POPULAR MECHANICS Magazine.

POPULAR MECHANICS Staff:

Editor-in-Chief: Joe Oldham
Managing Editor: Bill Hartford
Special Features Editor: Sheldon M. Gallager
Automotive Editor: Wade A. Hoyt, SAE
Home and Shop Editor: Steven Willson
Electronics Editor: Stephen A. Booth
Boating, Outdoors and Travel Editor: Timothy H. Cole
Science Editor: Dennis Eskow
Graphics Director: Bryan Canniff

POPULAR MECHANICS **Guide to Modern Telephones:**

Written and Edited by: C. Edward Cavert and George L. Hall
Contributing Writers/Consultants: Barrett S. Brandon, Jr., William Dare, George DesAutels, John Sabourin
Assistant Editor: Cynthia W. Lockhart
Manufacturing: Ron Schoenfeld
Production: Harriet Damon Shields & Associates
Art Director: Suzanne Bennett
Book Design: Suzanne Bennett & Associates
Illustrations: Vantage Art
Photography: C. Edward Cavert, Cynthia W. Lockhart, David Sailors, Charles Tack Photography (Reston, Virginia), Marianne Taylor
Photo Researcher: David Barkan

The staff of Hearst Direct Books is grateful to the following individuals and organizations:

Amidon Associates
Devoke Company
International Business Machines Corporation (IBM)
Inmac Corporation
Motorola, Inc.
NEC
Walker Telecommunications
Western Union

Technical Consultants:

Fred Langa
Clifford J. Pickering
Ron Schneiderman
Marc J. Stern

Picture Credits:

Reproduced with permission of AT&T Corporate Archive, page 1 (bottom left); page 6; page 11 (top left); page 14, page 17 (bottom right); page 19 (bottom right); page 20 (top right); page 23; page 24 (bottom right); page 35 (bottom left); page 36 (center right—left photo); page 39 (top right); page 73 (bottom right); page 75 (bottom right); page 102 (bottom left), page 121.

Bell Labs, page 1 (top, bottom center and bottom right); page 2; page 3; page 8 (top left); page 10 (top right); page 16; page 19 (center right); page 34 (bottom left); page 35 (bottom left, center left—right photo); page 36 (top left); color insert, page C1 (top).

Code-A-Phone, a Subsidiary of Conrac Corporation, page 39 (top left); page 57 (top right); page 58 (top right); page 59 (bottom left); page 107 (bottom left, top right).

Colorado Video, Inc., page 151.

Commodore Electronics, Ltd., page 46 (top right); page 63 (bottom right); page 67 (top right); page 112 (bottom right).

GTE Service Corporation, page 7; page 11 (bottom left); page 36 (center left, bottom right).

Hayes Microcomputer Products, Inc., page 64 (bottom left); page 70 (center left); page 113 (bottom right).

Image Data Corporation, page 149 (bottom right); page 150.

MCI, page 20 (bottom right).

Mitsubishi Electric Sales America, Inc., page 149 (bottom left, top right).

Northern Virginia Community College, page 148.

Panasonic, page 36 (center right—right photo); page 38 (bottom left, bottom right); page 39 (bottom left, bottom right).

Prometheus Products, Inc., page 34 (top right); page 64 (top left); page 65 (bottom right); page 69 (bottom right); page 70 (bottom left); page 71 (top left); page 112 (top right).

Radio Shack, a division of Tandy Corporation, page 11 (bottom right); page 13 (bottom right); page 24 (top left); page 35 (top right); page 36 (bottom left); page 37; page 40; page 41; page 46 (top, center and bottom left, bottom right); page 60 (bottom left, top right); page 64 (bottom right); page 98 (bottom left); page 141 (bottom right); color insert, page C1 (center, top and bottom right).

The Source, page 142.

Universal Data Systems, page 26 (bottom left).

Western Electric, page 146.

Library of Congress Cataloging in Publication Data
Main entry under title:

Popular mechanics guide to modern telephones.
Includes index.
1. Telephone—Amateurs' manuals. 1. Title: Guide to modern telephones.
TK9951.P67 621.386 86-25798

ISBN 0-87851-092-3

10 9 8 7 6 5 4 3 2 1

Manufactured in the United States of America.

Although every effort has been made to ensure the accuracy and completeness of the information in this book, Hearst Direct Books and Popular Mechanics make no guarantees, stated or implied, nor will they be liable in the event of misinterpretation or human error made by the reader or for any typographical errors that may appear. WORK SAFELY WITH HAND TOOLS. WEAR SAFETY GOGGLES. READ MANUFACTURER'S INSTRUCTIONS AND WARNINGS FOR ALL PRODUCTS.

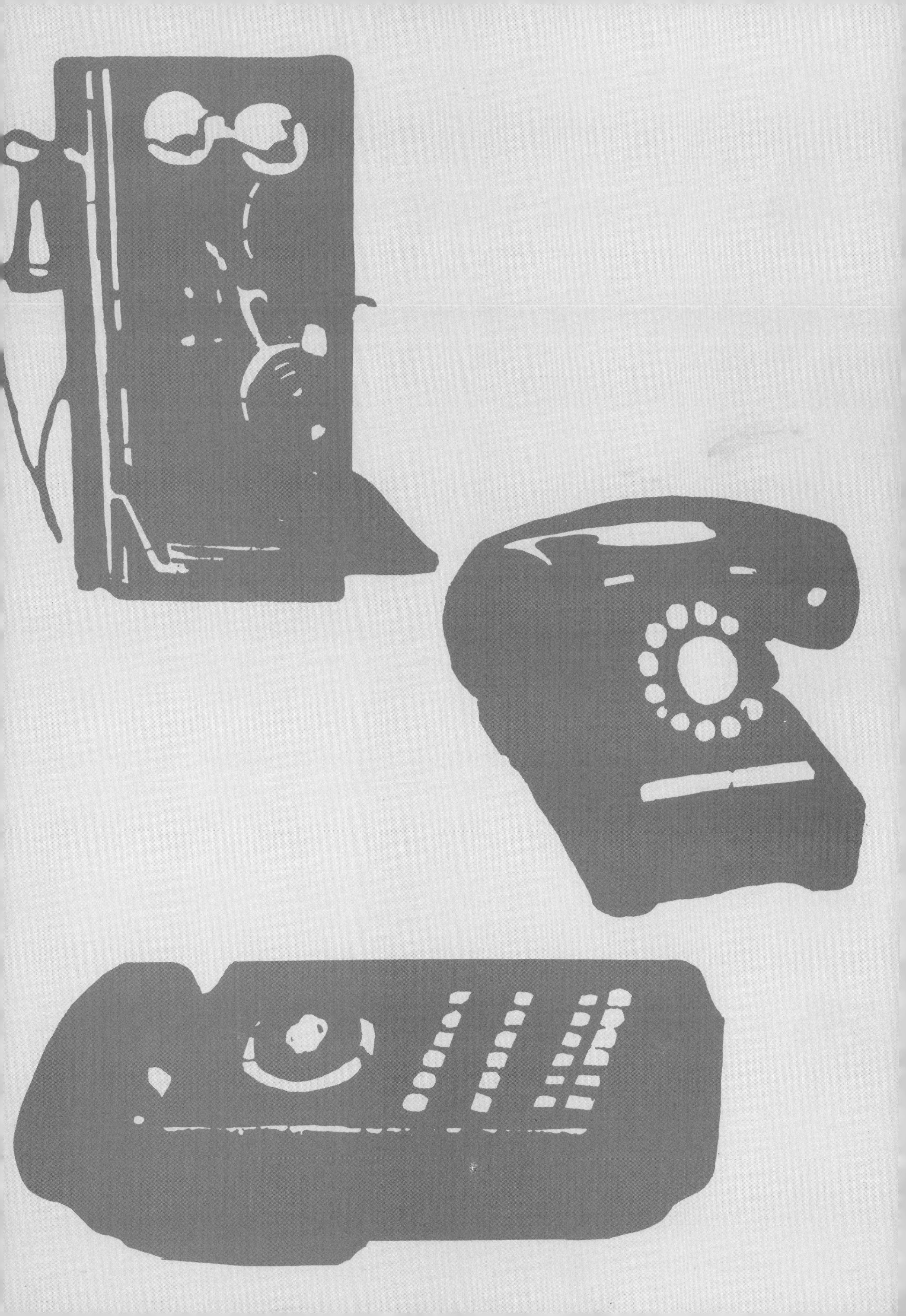

Contents

1 Who's in Charge Here?

In many ways deregulation and divestiture in the phone industry have had a profound effect on the consumer.

The Good Old Days

Twenty years ago, all telephone equipment in the home, including the wires and connectors, was owned by a local telephone company. As a customer, you rented the apparatus as part of your local and long-distance telephone service. Since the equipment belonged to the company, installation and maintenance were also provided as part of the service. While there were usually extra charges for installation, repairs were normally done by the company without direct cost to you.

In those "regulated" days, you were not allowed to buy, install or repair any telephone equipment of your own. In fact, you could move telephone instruments around only as far as the company-installed cables could reach. Even extension cords had to be rented from the local telephone company.

Telephone service was considered a public utility like water and power. The corporations that provided the services were closely regulated by federal and state governments.

Ma Bell and the AT&T Giant

About 80% of the telephones in use were owned by the gigantic Bell system. The American Telephone and Telegraph Company, which was owned by the Bell system, provided long-distance services to all local phone companies. No other long-distance services were allowed anywhere in the country. Both Bell and non-Bell companies connected their local customers to AT&T's long-lines.

The manufacturing arm of the old Bell system was the Western Electric Company. It turned out the telephone instruments that Bell installers placed in the homes and businesses of Bell subscribers.

There were (and still are) local telephone companies other than those owned by Bell (such as Central Telephone, serving the rural area around Aurora, Nebraska, and Continental Telephone in Manassas, Virginia). Many of these had to tie into Bell systems in neighboring communities much like a long-distance service.

Western Electric was not the only maker of phones. General Telephone, Kellog and Automatic Electric (among others) made the phones

The style and color of telephones you rented from the phone company in the days before deregulation were limited to the basic black cradle phone with a rotary dial and usually no long extension handset cord.

For many years, service to 80% of the telephones in the country was provided by local companies of the Bell system. AT&T long-lines provided long-distance service to both Bell and non-Bell companies, and most phones were made by Bell's manufacturing arm, Western Electric.

for the local non-Bell companies. The history of many of these companies goes back almost as far as the Bell system. The company that was later to become GTE Automatic Electric began phone service in 1889. It became the nation's largest independent telephone company outside the Bell system.

Revenue Sharing

During those times, it was national policy for the enormous revenues generated by long-distance calling to be shared by AT&T with other companies to reduce the cost of local telephone services in homes and businesses. In fact, home telephone costs in those days were very low — except for long-distance calling.

Business costs were somewhat higher, although AT&T did offer large discounts to firms that contracted for huge numbers of long-distance circuits. Few business subscribers and no residential subscribers, however, were large enough to benefit from these savings. For the most part, America's telephone service was notably better — and cheaper — than that in most other parts of the world. But there were problems.

Business Users Complain

Heavy long-distance users — especially businesses — were often irritated at having to subsidize local phone services. This complaint grew as business chains came to depend more and more on modern, nationwide telecommunications links between their many branch operations. This was especially true of firms and governmental units that made heavy use of telephone lines to connect their computers. In effect, large organizations used AT&T circuits, purchased in bulk, to set up their own private networks.

Then many users began to feel that the telephone utilities were not providing the up-to-date instruments and services that modern technology made possible. They felt that the lack of competition was beginning to make America's vast telephone system unresponsive and obsolete.

The Breakup of the Bell System

After some legal skirmishing, the federal courts began to agree with the critics. They reversed the long-time national policy of treating the telephone companies strictly as monopolies.

At first, the changes only allowed customers to buy approved telephone instruments and to connect them to company lines. However, this introduced competition into the design and sale of telephones and related equipment, to which the telephone utilities responded by expanding their own equipment offerings to new types and designs. The success of this new approach led the courts to try other kinds of deregulation as well.

New Long-Distance Services

New firms came into being that could lease massive numbers of long-distance circuits from AT&T at sizable business discounts and then resell the long-distance service at significant savings.

These organizations did not share their revenues with the utilities to reduce the cost of local telephone service. In light of this, the telephone utilities argued that this would have the effect of increasing overall customer costs, except perhaps for large-scale corporations.

After legal proceedings that lasted for a number of years, the federal courts decided that the objections by the utilities were not valid. They also decided that there should be more deregulation and not less. The courts decided that the Bell system was too big and powerful to operate competitively in the marketplace.

The Deregulation Decree

The historic court decree that required that

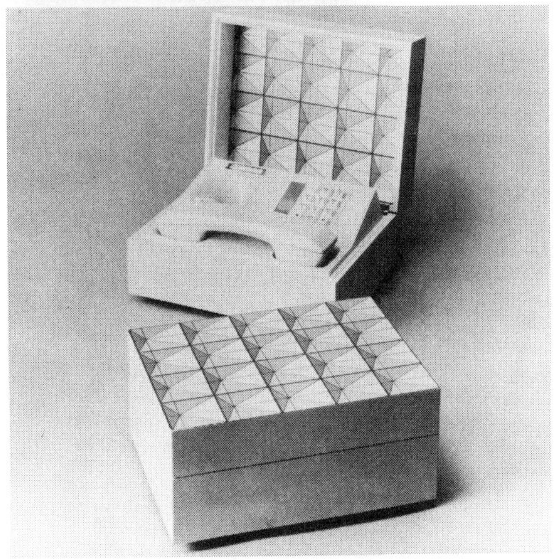

Exotic styles and shapes were introduced by other companies at the start of deregulation to compete with the plain old black telephone we had known so long. Inside the often bizarre cases, phones with advanced electronic technology made even the smallest entrepreneur competitive in the marketplace.

AT&T divest itself of the local companies in the Bell system occurred in January 1982. It also ordered that the local companies could no longer favor AT&T in providing long-distance service. And, most importantly for you as a home user, it insisted that the market for telephone equipment should be fully opened to competition.

But the court order did not totally end all utility regulation in the new telephone industry. It left the local telephone companies with a monopoly on lines (i.e., the circuits that make up the complex telephone networks) as well as the associated switching exchanges. This means that all telephone users must still order their outside lines and dial tone from a regulated utility organization.

The important exception to telephone deregulation is a logical one. If a community were to allow any number of separate firms selling lines and dial tones to operate, the criss-crossing of physical wires along streets and highways would be a nightmare. Moreover, you would likely have to order different lines from dozens of competing sources. Even in your home, you might have to have three or more telephones, each connected to a different exchange and long-distance service. Chaos!

Quality Controls and Regulated Standards

But the new national policy continued a provision announced when telephone customers

Separate wires used to be added for each new phone ordered, making telephone poles a major factor in large cities, as this early picture along New York's Broadway shows. Long-distance land-line service had to be installed and maintained through sparsely populated areas (such as this stretch through Nevada) for transcontinental coast-to-coast telephone service.

were first permitted to connect their own non-company equipment to utility-owned lines.

FCC Approval and Certification. Any apparatus connected into the telephone network has to receive an appropriate technical approval from the Federal Communications Commission (FCC) in Washington, D.C. This is to ensure that the apparatus is safe to use and will work properly within the network. It is illegal to sell telephone equipment for network use that has not been tested and approved to conform to FCC standards.

Even so, it is wise to look for the legally required notice on any instrument that will be physically linked into the telephone system in your home. The list of items includes telephone sets, telephone recorders, modems and cordless phones. The list does not include such items as wires and connectors.

Usually the FCC certification notice is actually printed on the casing of the equipment itself. Often, however, the notice also appears on the packaging label. The FCC certification itself consists of a series of letters and numbers following the phrase "Complies with Part 68 FCC Rules: FCC # . . . ." (This is followed by a long series of numbers and letters assigned to the manufacturer for that phone by the FCC.)

Proper Notification of REN. There is also a *r*inger *e*quivalence *n*umber (REN). The REN is needed by the utility in order for the utility engineers to be able to compute the precise electrical properties of the particular line on which it is to be used. This helps the utility check and fix any line problems that may arise.

You might be interested to know what information the REN gives the phone people. It indicates how close the electrical properties of the phone's "ringer device" are to those of a traditional telephone bell and on which side of the line loop it has been placed. It also tells them if your telephone system is drawing too much current, interfering with the phones of other subscribers. For you, the REN tells you how many phones you can have in your system. A good rule of thumb to follow is that the REN total of all your phones should not exceed 5.0; the REN of an individual phone ranges from about 0.4 to 1.3. (Many modern phones have ringer devices that are quite different from the old-fashioned telephone bell. Some even play chime tunes! Their electrical characteristics are correspondingly varied.)

By using information about your phones that you must report to the phone company, problems can be spotted without a technician's having to scrutinize every physical inch of the line itself.

Your Legal Obligation. You have a positive legal obligation when attaching any equipment to company lines (or to wires linked into these lines). As a subscriber, you must register the apparatus with the local telephone utility. You must also furnish the official FCC certification and REN of the telephone sets (or other equipment) you will be using. You can do this in writing or by calling the local company. You must always indicate the particular telephone line (by its directory number) on which the device is to be used. If you fail to notify the utility of the FCC certification and REN, the company has the right to cancel line service. The utility is not allowed to charge for processing this information or for letting you attach your own equipment.

If the phone company determines that your phone is causing problems with the telephone lines, you must disconnect the phone until necessary repairs (or replacement) have been made. If you don't, the phone company can cut off your service. The phone people should tell you in advance if your service is going to be cut off, but it may not always be practical for them to wait if the interference is too bad. They

COMPLIES WITH PART 68, FCC RULES
FCC REG. NO AS493N-70140-TE-T
RINGER EQUIVALENCE 0.8A
LISTED UNDER REA PE-41 1978

A certification number like this on any equipment connected to phone lines shows that it complies with technical standards set by the FCC. Without the certification, the equipment cannot be used legally.

COMPLIES WITH PART 68, FCC RULES
FCC REG. NO AS493N-70140-TE-T
RINGER EQUIVALENCE 0.8A
LISTED UNDER REA PE-41 1978

The ringer equivalence number (REN) must be on any telephone, answering machine or other device you connect to the phone lines. You have a legal obligation to register both the FCC certification number and the REN with your local telephone company.

should also tell you that you have a right, under FCC rules, to appeal this decision to the FCC if you feel you and your equipment are not the source of the trouble. Be prepared to prove your case if you appeal!

Phone companies can make changes in their equipment that could affect your home phone system. If you have special approved equipment on the line, like some computer modems, the phone company is required to give you written notice of any change they want to make that might affect the operation of this equipment. Then you can make changes in your equipment to allow for uninterrupted service. This is another good reason to follow the rules and notify the phone company when you put any new phone equipment on-line.

Incidentally, if you still have original Western Electric equipment supplied by the local utility company (or later sold or leased by AT&T) you don't have to provide the REN and FCC certification. This is because it wasn't the law when these older instruments were manufactured. Anyway, your phone company already has technical records about these phones.

The phone company's obligation for installing service to your home ends at a terminal like this. It is your responsibility to run lines from this terminal to the locations in your home where you want your phones.

Don't tamper with this grounding connection made by the phone company when they install your lines. The connection not only protects the incoming line and your phone system, but your phones will not work properly without proper grounding of the lines.

Who's in Charge of Home Wiring?

What about telephone wires inside a house or apartment that were originally installed by the local utility company? Who owns these nowadays? The owner of the building has inherited the use of them. The old wires legally still belong to the phone company.

Now, the utility company's own facilities end in a telephone terminal box (called a *protector*) in which inside wiring can be attached to a network line. The protector has been carefully grounded by the utility installer. Don't tamper with that ground arrangement or with the taps terminating the incoming telephone line. They have been fused to protect against such hazards as lightning. The three empty taps to which the house telephone wires are to be attached are plainly visible. These are the only ones you are allowed to touch. On many of the newer phone company terminals there is a modular jack into which you connect your home telephone wiring system. This saves you the trouble of hooking up wires to terminal screws. It also saves the phone company any concern about your disconnecting their wires.

Home Equipment

Under deregulation, the local phone company is no longer obligated to install or maintain equipment in your home or apartment. On the other hand, you can hire the utility to install and maintain the wiring (not the phone equipment, however), but the costs for these special services are not cheap!

Selecting and purchasing telephone equipment to be used in the home is now entirely your responsibility as a consumer. The utility is not really involved in providing and maintaining anything but local telephone lines and dial tone.

There are some exceptions to this for certain business customers as well as for residents served by independent (non-Bell) telephone utilities. (Other exceptions may involve the sale of telephone instruments by corporations affiliated with the Bell utilities. The new regulations in this area are legally complicated.)

Shortly after the legal breakup of the old Bell monopoly, its customers were given the option of buying or renting their on-premises Western Electric telephone set. You buy or rent these older phones from the new AT&T. Any required repair work is done — for a fee — by AT&T instead of the local telephone utility, as was the case "in the old days."

Long-Distance Services

Even the old concept of the network line itself has undergone some legal changes in the long-distance field.

No longer does AT&T hold a monopoly on long-distance services or facilities. Other enterprises are now encouraged to compete by building their own circuit systems (with satellites or microwaves), which can be leased for long-distance services to consumers, both large and small. Where desired, the new firms can even tie these new circuit systems into the older AT&T network so that long-distance calls can be switched to any telephone anywhere.

Communications satellites in orbit above the Earth replace the miles of wires strung along telephone poles linking your phone to any other telephone in transcontinental and international service by most long-distance companies.

You must now tell your local utility company which of the numerous long-distance suppliers you wish to use regularly. The network each consumer designates is then accessed through the local telephone by use of the familiar "1 + Area Code + Number" sequence. (Any other long-distance supplier can still be used from the same local telephone but only after the caller dials in a string of special numbers.)

In spite of all these historic changes, the huge AT&T network across the country remains the largest telephone system in existence.

Ironically, the AT&T telephone network itself remains a largely regulated utility, although the AT&T corporation is now free to engage in other forms of commercial initiative. For example, it is fast becoming a leader in the microcomputer field, an activity forbidden to it during its days as a cornerstone of the Bell system. In those times, Bell component companies could engage in nothing but telephone business. No sidelines were permitted for fear they would drive telephone costs up or exploit the massive capital amassed from the associated telephone monopoly to gain an unfair competitive advantage.

Where Is Deregulation Heading?

Where is deregulation heading? The hope is that we will have the best, biggest, most open and least expensive telecommunications system in the world.

For the residential telephone customer, the outlook is positive — especially if you can do the planning, purchasing, installation and maintenance without having to hire outsiders to do anything but bring a line with dial tone to a terminal box on your premises.

This book is designed to help you understand how telephones work, give you guidance in making your purchase decisions, help you install and maintain your phone equipment and give you a glimpse into some of the more advanced applications of our telecommunications system.

2
Mr. Bell's Invention

The telephone system we all know is both simple and complicated. The large-scale networks of circuits and exchanges that tie customers and communities together are telecommunications enterprises of great sophistication and complexity. The ordinary telephone instrument — which most of us call a "phone" — is an ingeniously simple piece of technology.

Making a Call

When you pick up the handset of your telephone to make a call, your telephone signals the phone company's branch in your neighborhood to send you a dial tone. You've made your first connection into the vast worldwide telephone network system.

The phone company also feeds your phone 48 volts of direct current to provide the voltage to make the microphone and speaker in the handset work.

As you dial the number, you send a coded signal for the seven-digit address of the phone you want to reach. (You may have to add three more digits for the code of the area of the country in which the phone is located if you're calling long distance.) Rotary-dial phones use electrical pulses to represent the number dialed. Electronic-tone dials use tone coding to indicate the number dialed. Either number code selects a route to a specific pair of wires connected to the one phone you want to call — anywhere in the country (or world-wide).

Behind that deceptively simple device you know as a telephone is a world of electronic and space-age technology standing by to link your phone to any other in the world.

If the phone you're calling is in use ("off-hook"), the phone company sends you a busy signal — a tone with a distinctive rhythm pattern telling you the phone is busy and to hang up and try again later. If the phone you're calling is not being used ("on-hook"), the phone company sends a high-voltage signal to make that phone ring. It also sends you a different signal in the same ringing rhythm pattern to let you know the other phone is ringing.

When the phone you're trying to reach is answered, the local exchange immediately stops sending it the ring signal. Your phone and the one you're calling then go to work, using the phone company supplied voltage to convert the sounds of voices into varying electrical impulses and back again to sound. These impulses are massaged by phone company circuits so that they can be transmitted back and forth along the entire network of circuits and lines that connect the two phones.

When you've finished your call and both parties hang up, the routing circuits between the local exchanges are cleared to be used for other calls. The phone company sends your on-hook phone a special signal that tells the utility you don't want to make a call as long as that signal is not interrupted (that is, until you take the receiver off the hook again). Your phone is now set to receive a ringing signal when someone calls you.

The complex technology behind this process we all have come to accept as a simple phone call has developed over the years from very humble beginnings.

Mr. Bell's Invention

The man credited with inventing and patenting the first practical telephone, Alexander Graham Bell, said it well at the very beginning. "If I could make a current of electricity vary in intensity precisely as the air varies in density during the production of sound, I should be able to transmit speech telegraphically." And that is precisely what he did back on March 10, 1876.

Bell developed a device that passed a low-voltage current through a primitive form of microphone. Airborne sound waves striking the thin metal diaphragm of the little microphone (or transmitter) caused the current to be varied — or modulated — in proportion to the strength

As legend has it, the first words were transmitted by telephone on March 10, 1876, when Alexander Graham Bell summoned his assistant, Mr. Watson, after Bell had spilled acid on himself.

of the sound itself. When the modulated signal was fed over a wire to a miniature speaker (or receiver), it reproduced the original sounds by sending the modulated current through a tiny electromagnet that vibrated a thin metal diaphragm. The speaker hasn't changed much since Bell's original design; microphones have changed, modifying Bell's principle with newer technology.

Microphones

Originally, Bell had used liquid acid as the modulating medium. In fact, legend has it that the first words actually transmitted by telephone, "Mr. Watson come here, I need you," were uttered by Bell when he spilled some of the acid on himself.

Carbon Mikes. The company Bell founded soon replaced the acid with a little package of carbon granules in the transmitter. It was found that the more densely these tiny grains of carbon were packed together, the more electricity could be made to flow through them. Consequently the Bell engineers placed the diaphragm of their simple microphone so that it could apply varying sound pressures to the carbon granules. The changing sound patterns varied how densely the carbon granules were packed together, which modulated the passing current.

Carbon mikes stayed with us for quite a while. (In fact they are still the most common type in use only because there are so many older phones in use.) But a way had to be found to increase the fidelity of the sound — and to make the mikes less expensive to build.

Dynamic Mikes. Instead of compressing bits of carbon to control the strength of electrical impulses, engineers found they could do this modulation by letting the sound vibrate a metal diaphragm placed in an electromagnetic field. The more the diaphragm moved in the field, the stronger the current that was sent out. This electrodynamic microphone sounded better, but it wasn't much cheaper to build.

Electret Mikes. Most new phones today are built with electret microphones. The pickup in an electret mike is one plate of a capacitor that goes to work directly on the voltage of the circuit when the plate is moved by the sound waves. The dielectric used in this mike stores electricity almost indefinitely, eliminating the need for heavy, expensive and current-hungry electromagnet coils. The package of carbon granules is now a relic of the past.

The Telephone Handset

The pickup and speaker circuits in the handset

Bell's first telephone. His invention converted the sound waves of a person's voice into electrical impulses that could be sent along wires to a similar device hooked to the other end.

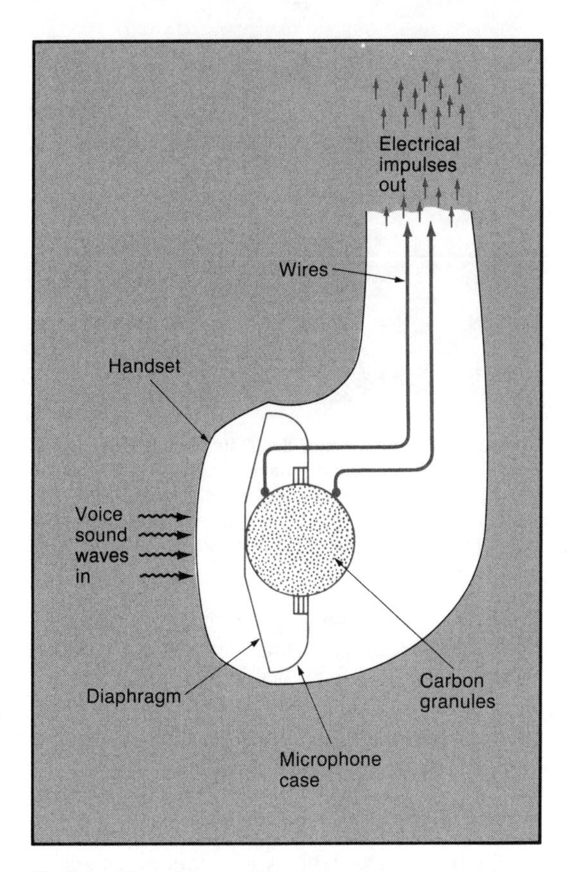

Carbon microphone. A thin diaphragm is connected to a package of tiny granules of carbon. As sound waves move the diaphragm, the electrical characteristics of the carbon granules change, sending a signal that varies in proportion to the tone and loudness of the sound waves.

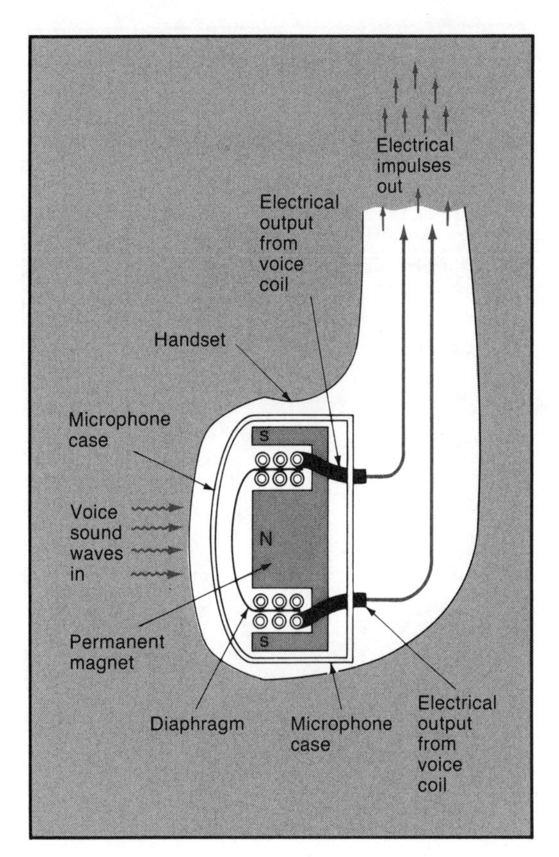

Dynamic microphone. The diaphragm is in an electromagnetic field. When sound waves cause the diaphragm to move, electrical current is induced in the coil of wires around the diaphragm, creating the varying electrical impulses that are sent along the wires.

of a phone are isolated from each other so that the phone does not try to convert electrical impulses back into sound in the microphone instead of the speaker.

Sidetone and Speaking Volume. In the early days, if people shouted too loudly, their audio became distorted, or the loudness interfered with calls others were making on the same circuits. Engineers found if they sent a little bit of your voice back into the receiver to let you hear yourself talk, you would know the system was working and would not shout to be heard. Telephone people call this "sidetone," and engineers have developed an exact balance of volume; too little sidetone and people still shout; too much sidetone and they don't speak loudly enough to be heard.

Telephone Fidelity. The fidelity of telephone audio is extremely limited. Its bandwidth of some 3000 Hz is about as narrow as the human ear will allow to distinguish speech sounds. This stingy attribute actually makes our telephone circuits more efficient because it allows more calls to travel on wires, microwave beams or satellite channels.

A Safe Home Appliance

The use of low-voltage, low-current electricity — usually 48 volts at about 1/10 amp — makes the telephone a safe technology to have around the home. The signal to make your phone ring is a little higher voltage (about 90–120 volts) but the current is likely to give you just a good tingle if your fingers are in the wrong place when your phone rings. The same level of ordinary AC voltage in the power outlets of your home is more likely to send you flying with a good zap because of the high current it supplies.

Reaching Out

At the very beginning of the telephone age, the

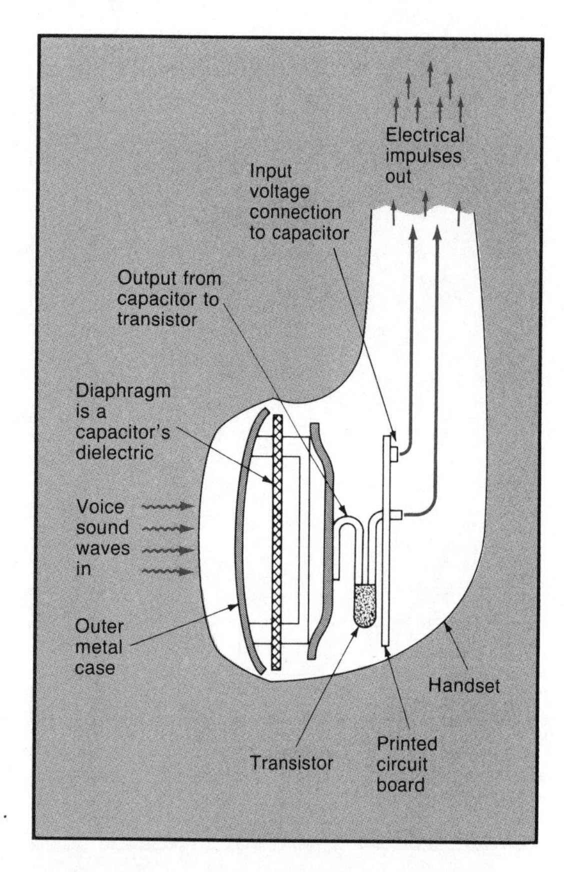

Eletret microphone. The diaphragm is one plate of a capacitor. As sound waves move the diaphragm, the capacitance value of the device changes, varying the electrical output. Much higher voice fidelity and lower manufacturing costs are possible with the mike.

connecting wires ran directly from instrument to instrument. The utter impracticality of such an arrangement became apparent very quickly. Some sort of "hub" switching system had to be developed to route calls from phone to phone.

This gave birth to the telephone exchange in 1877. An exchange is a complex arrangement of relays and electronics for routing telephone calls from line to line.

Operator-Assisted Calls

In the early days, the connections were made manually by operators. These switchboards had separate jacks that went to each phone or to other local exchanges. To place a call, you had to request the connection by speaking to the operator over the phone. Then the operator inserted the plug with your incoming call into the requested jack on the switchboard. The operator then keyed a special "voltage interrupt" signal to ring the requested phone's bell. Once the

To route calls to their final destination, early telephone exchanges used electromechanical relays and devices known as "steppers" connected by miles of wires in large bundles. Almost all of these have been replaced now by electronic switching centers.

connection was completed, the operator disconnected from your line and waited for the next call to come in. It was a slow procedure at best. Long-distance calls often went through many operators, each doing the same procedure to move your call along.

Rotary Dials

At the turn of the century a telephone engineer named A. B. Strowger developed a much better way of exchanging calls: the dial telephone. Dialing permitted you to signal the desired number without an operator on the line. You picked up a phone, listened for the buzzing dial tone on the line (fed in by the telephone exchange) and then dialed the required sequence of numbers, one by one, on a rotary-dial device built into the instrument.

Each number (or letter) dialed caused the outgoing line voltage to be pulsed on and off in a precise timing sequence, which could be detected by electrical sending equipment at the exchange. The pulses were timed by a governor on the dial mechanism so that the interruption of line voltage was long enough to signal circuit routing, but not long enough to make the system think you had hung up.

These dial pulses were able to trigger a series of mechanical (or electrical) relays, which brought the required lines into physical contact, allowing the call to go through to the indicated telephone. The ring signal accompanied the call automatically if the phone you were calling was not in use.

The dial exchange replaced the operator virtually everywhere but in the long-distance sys-

Operators routed calls on switchboards like this in the days before dial phones. Every telephone in an operator's small area terminated in a jack on the switchboard. The switchboard also had jacks connected to other boards and to long-distance "trunk" lines. Today's operators not only have more modern equipment, they now handle only special calls requiring operator assistance like collect or person-to-person calls.

tem. There were too many dial pulses and relays involved to route long-distance calls to the one specific final destination telephone.

Electronic-Tone Dialing

The Bell system's Touch-Tone™ development turned the corner on this problem. Telephone people call this system DTMF, which stands for *d*ual-*t*one *m*ulti*f*requency. It works this way: The telephone instrument contains a small, complex system of audio frequency generators. This system produces different audible frequency tones. Seven of these tones are used to dial. When you press any button on the phone's built-in keypad, a different combination of two of the seven tones is generated. Each tone is of equal electrical strength, and they start and stop at the same time. The seven individual tones are arranged in a matrix so that their dual combinations allow twelve distinctive tone pairs to be triggered by the individual keys for dialing.

The tone generator actually produces eight tones for a total of sixteen dual-tone combinations. The other tones are there for future expansion by the phone companies. That way, equipment now in use won't be obsolete when they want to expand the kind of signals you send out. (It was also easier to build the tone generators that way.) Some business applications now use the four extra dual tones for internal office communications and other special purposes.

Dial phones let you connect directly to another phone without an operator's help. The rotating dial interrupted the line voltage in a carefully timed sequence, triggering a series of relays at the local exchanges to route your call to just the one pair of wires connected to the phone you were calling. In early rotary-dial service, operators were still needed to call long distance. Operators could also route the calls, helping those for whom the new dialing procedures were too complex to cope with.

Most telephones today use precisely tuned electronic tones to dial. These tones signal the integrated circuit components in local exchanges and electronically route your call to its final destination. In many phone companies, connections between local exchanges and connections to long-distance services are now by electronic tone, even though you may still have rotary-dial service to your local exchange.

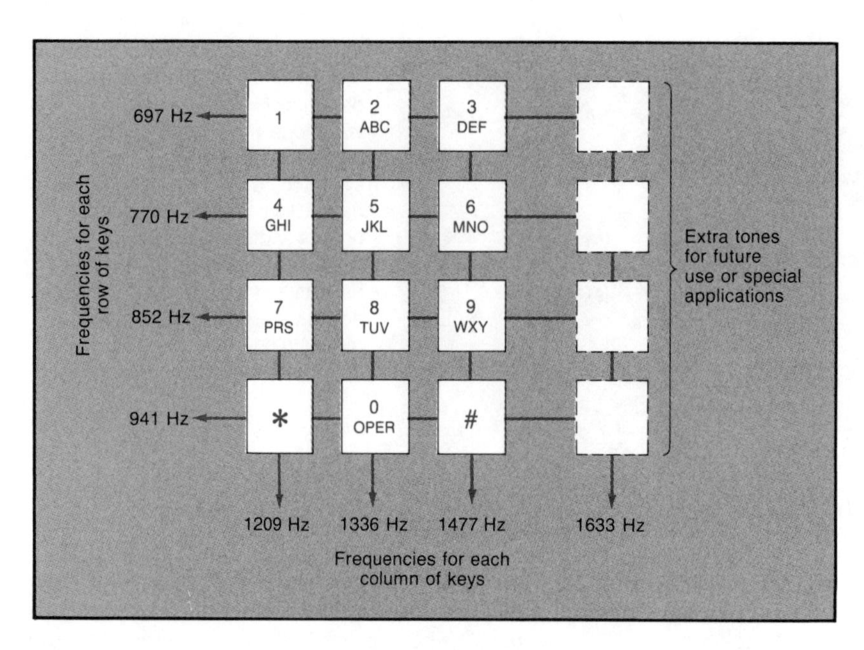

Electronic-tone dialing is coded by the frequency of each of the tones sent out. A set of tones at one frequency is sent out for each *row* of keys. Another tone corresponds to the *column* the key is on. If you pressed the key marked "6," for example, one tone at 770 Hz and another at 1477 Hz would be sent out. Equipment everywhere will know which key has been pressed because these two tones can only mean one key in the matrix. Almost all phones have sixteen distinct dual-tone patterns, although only twelve are used on the conventional dial. The other four are reserved for future use or for special applications for which your phone is wired, such as automatic dialing or redial.

The asterisk (*) and number sign (#) on tone dial pads are also put there to use in activating special features such as speed calling and conference calls.

Using the Dial Codes

The dial system, tone or pulse, is used to select the routing along the telephone circuits that lead to just the one phone you want to reach. The dial system reacts to only as many numbers (or letters) as are needed to complete the routing of a call. Additional pulses or tones are ignored by the phone companies.

Pulse Dial Routing. Pulse dialing simply turns the line voltage on or off in the coded sequence. The rotary-dialing signals become inoperative once the local exchange relays have connected the lines to complete the call. Additional dialing will interrupt the voltage (and produce irritating sounds in the ear) but it won't trigger any additional relays at the exchange.

Dual-Tone Routing. Dual tones travel along the line to an exchange where electronic filters detect which keys produced the sounds and carry out the relay switching accordingly. You can continue to send dual tones over the lines without interrupting the line current as pulse dials do once the phone companies have completed the connection. (Some people have even worked out simple electronic songs they can play to their friends during a call by tapping in a sequence of keys.)

Added Tone Services. The tone system is also what makes it possible for a telephone to be used to activate such services as computer banking.

The tone phone keypad becomes a miniature keyboard with each key representing a number, letter or symbol. After being connected to the bank's computer, for example, touching the keys allows you to type in various instructions and information, which the circuits in the service can act on. Since the phone company does not respond to these tones, they can be used by telecommunications services (like electronic banking) any way they choose.

Pulse versus Tone

Tone dialing is not only more versatile than

pulse dialing, it is also notably faster — both for the caller and at the exchange. Virtually all local exchanges now can handle electronic tones. These same exchanges will also respond to the older rotary-dial pulses.

Tone service usually adds a small extra charge to your monthly bill. With some of the newer phones available today, you can keep the less expensive pulse dial service and still access tone dial services you may want to call. Once the call is completed with the pulses your key pad sends out, you move a switch on the phone to change to tone signals and then send dual tones for electronic banking, telecomputing data retrieval or the host of other tone signal services emerging almost daily.

The dial pad on an electronic-tone phone has the numbers 1–2–3 along the top row, while virtually all calculators and adding machines have the numbers 7–8–9 along that row. For people used to entering numbers by touch on a calculator or adding machine, an adjustment was necessary. Most people, however, may never have noticed the difference.

The New Look in Phones

With tone switching came the new key telephones as well. Their keypad of twelve keys was easier and faster to use than the old rotary dial. (It is interesting to note, however, that the key arrangement on a telephone keypad is the reverse of the arrangement on a modern calculator or computer. There is no good reason why this is so!)

At first, there was no practical way for the average consumer to use the new key arrangement on pulse dial lines. But engineers soon came up with an instrument that had a keypad but that generated the older dial pulses.

These inexpensive "pulse pad" phones are easier to use than rotary-dial models but they can't be used as substitutes for true electronic-tone telephones. They even sound different to the user. As you press the keys, you hear the familiar rotary-dial pulses going out on the line. Sometimes pulse dialing with a keypad is disguised by a tonelike sound, but it is not true electronic DTMF dialing. The sound only tells you when the key is activated; it will not work where true electronic dial tones are needed.

Other Kinds of Tones

Your local exchange handles other tones besides the electronic dual tones for dialing. The dial tone you get when you pick up the handset to make a call is the most obvious.

The high voltage sent to make your phone ring is seen by the exchange and by telephones as a special kind of tone: it has a specific AC frequency and voltage.

Another AC signal you hear is the busy signals (there are two kinds). If you listen carefully, sometimes you'll hear the familiar busy signal sound but with a faster cadence. This tells you the circuits (usually long-distance) are congested or busy and your call can't get through to the local exchange of the number you're calling. That phone may not be busy at all.

The local exchange will also send you a very annoying tone if you've taken (or left) your phone off the hook without starting to dial. This tone gradually gets louder the longer the off-hook condition lasts, until you can hear it from almost anywhere in the house. Some phone companies will just give up on you after a couple of minutes and stop the tone; others will keep pestering you until you hang up the phone.

The Phone Is Ringing! (Chirping?)

In the old days, the jarring sound of a clapper

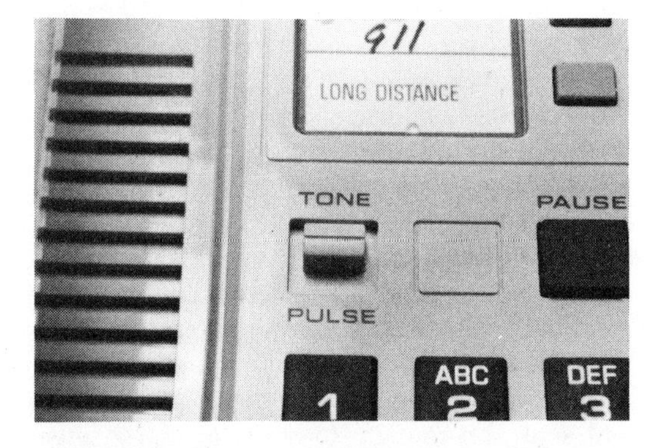

Many phones have a switch that lets you change from pulse dialing to tone dialing. This lets you dial into a long-distance service or information base using the more economical pulse service, and then switch to tone dialing to continue your own routing with the electronic signals the service requires.

slapping against two bells was a force to be reckoned with. When a phone rang, people answered — often ignoring social etiquette or business protocol. Now many telephones have abandoned the vigorous bell ringer for the tweets, chirps and warbles generated by electronic circuits.

The new sounds take some getting used to. You may not even realize your phone is ringing the first couple of days you change to new electronic phones. People with hearing loss in the higher frequencies may have trouble hearing some of these ringers at all.

Wires and Exchanges

At the start of our great telephone network system, signals were transmitted from one place to another on small copper wires strung on poles along streets and highways. The ratio of one call per wire was a poor one. Too many wires were needed. These pairs of copper wire encased in plastic sleeves and braided together are still what most of us think of as old-fashioned telephone lines.

At one time, each telephone call ran along separate wires strung on poles in our cities and across our countryside.

The Twisted-Pair Network

Electrical engineers soon figured out ways to "multiplex" several different signals simultaneously on the same wire. Multiplexing techniques have become more sophisticated over the intervening years. In the early days one pair of wires had to carry your voice to another phone; another pair carried that voice back to you. This four-wire system has been replaced by more modern transmission systems. The twisted-pair wires are still used now to connect your home to the local exchange, but fiber-optic strands may soon replace even this link. Between exchanges, however, telephone companies and long-distance carriers now use analog or digital multiplexing to combine many voice signals on one transmission medium. In analog multiplexing, each call takes up a very small band of frequencies on the wires, coaxial cable, microwave, satellite or fiber-optic transmission medium. In digital multiplexing, your voice signals are converted to a continuous stream of on–off signals that are combined with others and sent very rapidly over a single transmission medium. Special circuits at the phone company convert the signals for either multiplexing method so they can be transmitted more efficiently but still not cause any problems with the microphone and speaker circuits in your phone.

Telephone engineers developed methods of burying the cables to protect them from hazards like ice and wind storms. In order to transmit the low voltage signals over long distances, special electronic amplifiers were developed. These made it possible to send telephone signals as far away as the other side of the Atlantic Ocean. Where cables couldn't go, the industry made use of radiotelephone connections.

Engineers also developed cables of greater call-carrying capacity. The best known of these is the familiar coaxial cable used also to feed CATV systems and to connect your VCR to the TV set. Modern telephone networks now make efficient use of microwaves, satellites and fiber-optic strands to carry massive numbers of calls between exchanges and over longer distances. In the home, traditional copper wiring is still the only choice, but even in the home the dominance of copper wire is not likely to last long. Special wide-band fiber-optic strands will be used soon in the link to your home and eventually even in the connections to individual telephones inside your home.

Party Lines

On a party line, you share the pair of wires

going to the local exchange with one or more of your neighbors. Special connections in your telephone and in the local exchange can send a ringing signal to only your phone on most party lines, but if someone else on your party line is making a call, you cannot receive a dial tone. In fact, when you pick up your phone to make a call, you'll hear the other people talking on the line.

If you have a party line, you will need special connections in your phone to receive the proper ring signal and not to interfere with the performance of other phones on your party line circuit. You cannot buy and use just any phone off the shelf. The phone will have to be modified by the store that sells it to you. This is not a task for the do-it-yourselfer.

Party lines were widely available, even in concentrated urban areas like New York and Chicago, when telephone systems were still in the building process. Now, however, this type of service is largely limited to rural areas or towns served by smaller telephone companies, where distances or type of terrain do not make it economical for phone companies to string separate wires to each home. You can't order party line service in most major metropolitan areas now even if you wanted to.

Home Wiring

Most home telephone wiring consists of four color-coded, plastic-covered copper wires encased in a plastic sleeve. Telephone people call this type of wire "quad." Interestingly, only three of the enclosed wires are customarily used in residential installations. The fourth is an extra to be used only in specialized situations as, for example, when current is needed at the phone to light up a phone dial or activate an intercom buzzer.

The three wires used in home settings are termed *tip, ring* and *ground.* The tip and ring wires are red and green, the ground wire is yellow and the extra is black. These rather strange names for the wires date back to the operator switchboard days of telephones. One wire was connected to the tip of the switchboard plug the operator used, another to a ring between the tip and the sleeve, and the third to the grounded sleeve.

There is also a six-wire cable that lets you run two separate lines inside your home. This cable has two ring-and-tip pairs with a common ground and one extra.

Wires running to and from the phone company's terminal in your home — called a *protector* — terminate at modular jacks on your walls. You plug the phones into the wall connectors using flexible cords with tiny modular male plugs attached to the end. (Older homes may have a four-prong connector arrangement. There are adaptors to convert these into the newer, smaller modular formats.)

If the phone's cord is too short to reach the wall connector, you can snap a simple extension between the plug and the connector. You can use various modular adapters to ensure proper match and fit. The whole modular wiring concept has been designed to make it extremely easy for you to run your own wires when and where you need them.

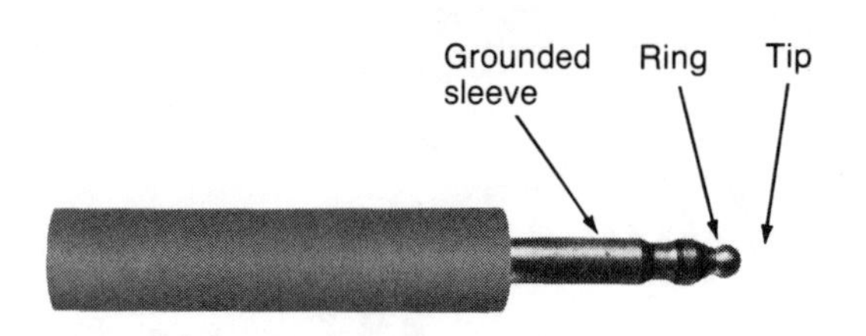

In early phone systems, one wire was connected to the tip of the switchboard plug. Another was connected to a ring on the plug just below the tip. The third was connected to the grounded sleeve. This is why the three wires you connect in your home telephone system are called "ring," "tip" and "ground" by phone people.

Virtually every telephone and telephone accessory made now comes ready to connect to the universal modular plug and jack system. Phones sold in the early days of deregulation may still have the older four-prong connectors; these are easily converted to modular with inexpensive adapters.

3
Your Connection to the World

When you dial or key in numbers on your home telephone, you are entering a vast electronic world. The dial pulses or tones your phone sends out on your line connect you to points almost anywhere on earth.

THE NEIGHBORHOOD EXCHANGE

The signals first have to go to a special facility in your own neighborhood. It's called a telephone exchange. You might never have noticed it from the outside, because it's likely to resemble a perfectly ordinary building. No big signs are apt to strike your eye.

If you could see what's inside (and you're not likely to be invited in), you would be impressed with the "high tech" appearance. There are orderly racks (or frames) of miniaturized electronic or mechanical relays (called switchers) connected to thousands upon thousands of tiny wires. There are various kinds of switchers including steppers, crossbars and electronic. Their differences are not always important to the caller.

The local exchange is the facility that feeds the familiar dial tone out to your phone to let you know you are connected into the system.

How You Get a Dial Tone

Statistical computations are used to make the dial tone available in a ratio that dramatically

The local exchange, or "central office," in your neighborhood is deceptively ordinary on the outside.

High-tech apparatus and electronic circuitry fill the inside of the local exchange buildings. It is here that your call begins its journey along the interconnected system to its final destination, the number you've called.

reduces the number of active lines connected to the exchange switchers at any given moment. The systems works this way.

When your new telephone line is installed, your likely calling patterns are statistically projected on the basis of such things as your location, type of residence, family size and type of service.

Then you're assigned to a suitable *hundreds group* at your local exchange. Because of the calling patterns of your group, the local exchange needs to feed dial tones to only about 10 of the 100 lines at any given time. The other 90 lines are quiet and unused.

The telephone companies have learned over the years that calling patterns are reliably predictable. This arrangement saves them the cost of activating all the lines simultaneously.

Of course, statistical projections are far from perfect. The utility maintains equipment that can override the hundreds group when more than ten lines at once want to place a call. Even so, there may be a slight delay in getting dial tones out to the lines. In a time of emergency when lots of people are calling at once, these delays can be several seconds long. Fortunately, that problem rarely happens.

Call Routing

The local exchange is also the point at which your dialing information makes its first (and sometimes only) impact. It triggers relays leading to the circuit containing the numbered line you wish to reach, or to the next stop on the logical route to that line.

If the number is available in your own exchange, the call goes directly between your line and the cable leading out to the phone location you are calling. The little relays will have provided a direct electrical contact path for you.

The outgoing local cable — like the incoming one — contains a sizable number of different wire pairs (usually called loops) and is capable of handling a large number of separate calls simultaneously. The tones or pulses you've dialed will guide your call to the one pair of wires connected to the phone you wish to reach.

Call Multiplexing

Early on, telephone engineers realized that it was inefficient to use a copper wire to send only a single call at a time between relatively distant points. They developed a technique called *carrier transmission* to improve things.

It works this way. Suppose a given strand of wire is capable of physically transmitting 12,000 Hz (cycles per second) worth of information. Suppose a single telephone call consumes only 3000 Hz of this frequency bandwidth (although 4000 Hz is set aside for a voice channel to minimize interference). There is 8000 Hz left over — enough for two other calls.

Suppose you could stack up three calls so that the 4000 Hz bandwidth reserved for each call fell at different start–stop points inside the total 12,000 Hz bandwidth of the wire. Call 1 would use a center frequency, or carrier, of 2000 Hz with the signal modulated between 0 and 4000 Hz; the voice signal of call 2 would vary between 4001 and 8000 Hz; and call 3 from 8001 to 12,000 Hz around a carrier frequency of 10,000 Hz.

This is what happens with analog carrier transmission. Multiple calls are stacked up simultaneously around different carrier frequencies generated at the transmitter end. At the receiver end, special electronic filters convert the individual calls back to the voice bandwidth. For example, call 2 in the example above would be reconstituted from 4001–8000 Hz back within the 1–4000 Hz voice channel. Then, the individual calls are sent in their restored state along wires from the local exchange to the phones destined for them. There, the signals reproduce the original voice sounds that spawned the call only split seconds earlier.

Analog multiplexing has been replaced in most parts of the country by digital multiplexing. While the ways of stacking the different voice channels differ greatly with analog and digital multiplexing, the basic concept of combining calls on a single transmission medium illustrated here remains the same. Now, coaxial cable, fiber optics and wireless transmission channels can handle thousands of channels of voice communication.

Electronic circuits and relays route your call to just the one pair of wires leading to the home or business of the number you are calling.

Long-Distance Routing

If the call you had dialed was meant to go outside your own local exchange, the relays and switches would send your signals into a circuit bound for some other, more distant exchange. In fact, this forwarding process could occur several times as your call darted on toward its ultimate destination.

Your call may have a strange, often bizarre routing to its final destination. If you live in Virginia and call Ohio, your call might be routed through New York and Pennsylvania if all direct circuits are busy. It could even be sent by satellite to California and back by land line all the way to Ohio. The long-distance company's objective is to get your call through — and quickly so they can release their equipment for other

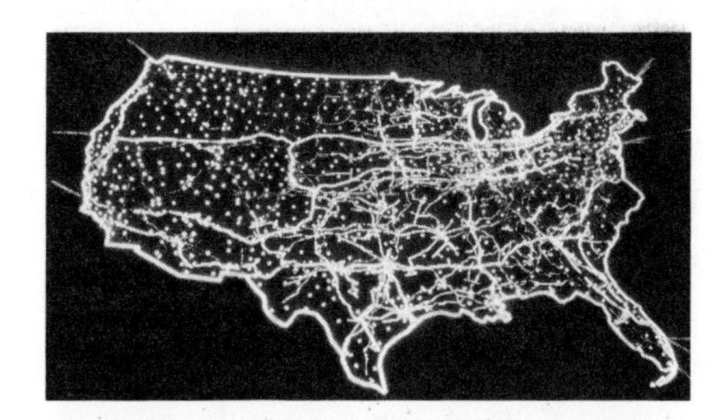

Long-distance calls between Los Angeles and Seattle through San Francisco can be routed over coaxial cable, microwave radio, satellite transmission or fiber-optic cable. Calls from the East Coast bound for Seattle could be routed through Los Angeles and San Francisco first if direct circuits are all in use.

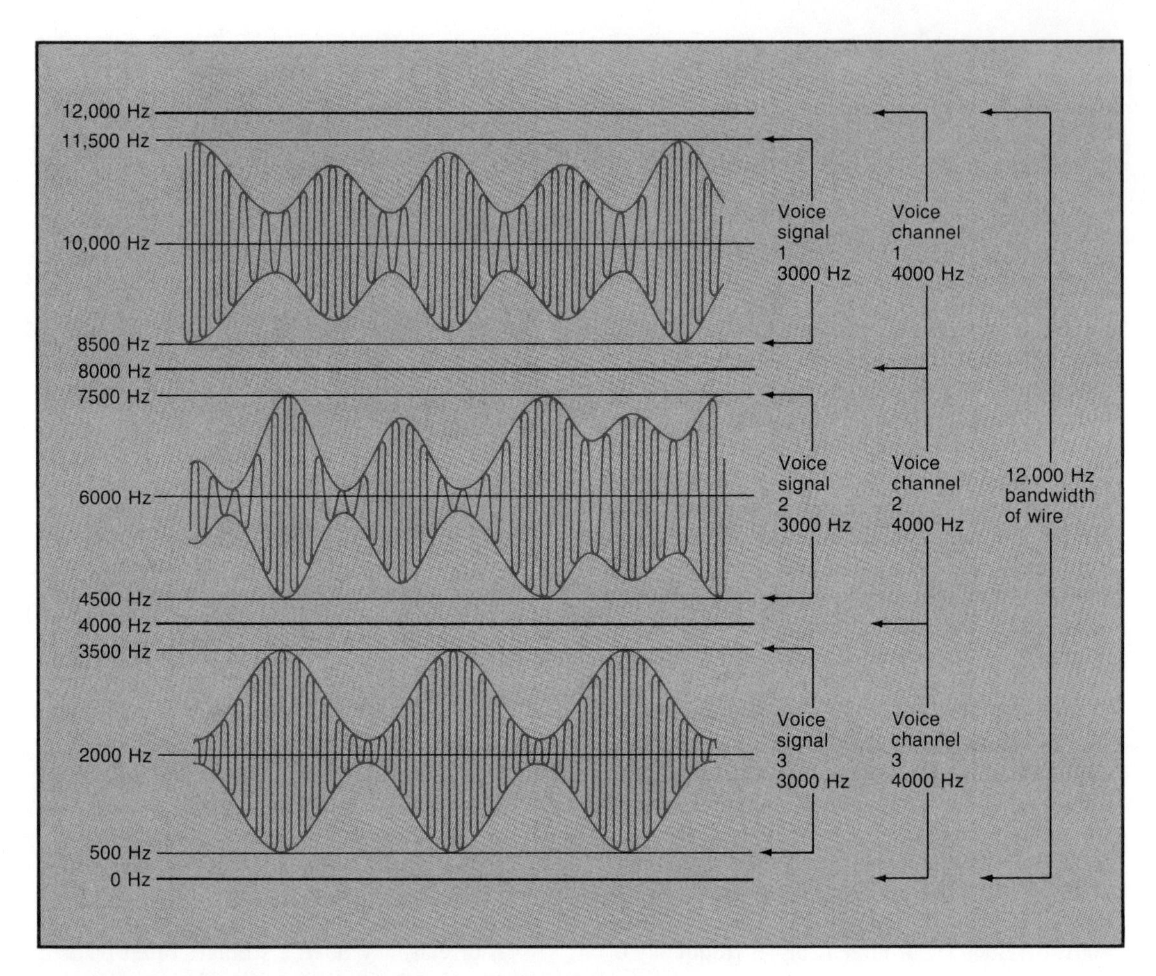

To increase the call-carrying capacity of any transmission channel, phone companies combine many calls or voice channels on a single wire or radio frequency. The process is called *multiplexing*.

This diagram illustrates how three calls are multiplexed on one wire. By using a different center frequency to modulate the voice signal of a call, each will fit in a voice channel using only 4000 of the 12,000-Hz bandwidth available. In this example, three calls can be sent over the same wire used for just one before. Before the calls reach your phone, the others on the wire are filtered out and you hear only the voice signal intended to reach your phone.

While frequency multiplexing is shown here to illustrate the concept, most multiplexing now is digital. While the actual bandwidth-sharing scheme is different in digital multiplexing than shown here, the basic concept is the same.

calls. They don't really care how far it has to go to get there. You are billed only for the distance between your phone and the one you called.

In some instances the call would alert a human operator to cut in to ask questions needed to complete the process. This would be true where you were calling long-distance person-to-person or where direct-dialing facilities were unavailable (if any such places still exist).

TRANSMISSION CHANNELS

The means used to relay your call signals from place to place might vary from traditional copper wire to coaxial cables to fiber-optic cables to microwaves to satellite channels. Sometimes all of these transmission paths might be used to complete a single call.

Coaxial Cable

The coaxial (or coax) cable is a flexible tube in which one conductor is a sleeve that completely encircles the other, which is a wire. The more expensive coax cable provides greater bandwidth capacities and greater resistance to external interference. Both improvements let coax cables carry many more calls than the twisted-pair copper wires.

Fiber Optic

Fiber optic is the most modern cable form. It doesn't use copper at all. It uses a highly specialized strand of bendable glass through which light rays can travel relatively long distances — even around corners. Telephone voice channels are modulated laser beams instead of modulated radio frequencies.

The capacity of this very expensive kind of cable is mind-boggling because it is so much greater than that of other cable types. Over 8000 voice messages can travel simultaneously on a fiber-optic "wire" no larger than a human hair. By the 1990s, advances in making the optic fiber and in the laser-driven transmission will increase the capacity fourfold.

Fiber optics is not only being used where high message concentrations occur, as for example, between New York and Washington, D.C., but nowadays even in remote rural areas, and soon even in your home.

Wireless Technologies

The wireless circuit technologies all involve radio-type electromagnetic transmissions through the air — or space.

Microwave and satellite relays are used to connect exchanges over the vast long-distance networks. These two differ largely in the distances they reach. Since the radio waves on the frequencies used tend to follow straight lines, the high-flying satellite systems far exceed the earth-bound microwave links in the distance required between the transmitting and receiving locations.

Microwave Links. To reach across country with microwave telephone links, AT&T had to

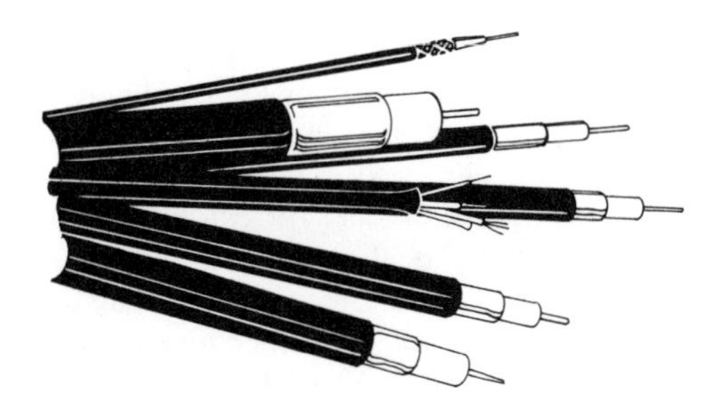

Coaxial cable is a flexible tube in which one conductor is a sleeve that encircles the other, which is a wire.

A single fiber-optic strand, no larger than a hair, is capable of carrying over 6000 simultaneous telephone conversations. The larger cable of twisted pairs of copper wire shown below the fiber-optic strand can carry only 2400 telephone conversations.

Radio transmission using the frequencies of the microwave band has replaced land-line wires in most long-distance links and transcontinental telephone service. It is common to see these microwave towers, filled with transmitting dishes and receiving reflectors, every 30 or 40 miles apart across the country.

Satellite dishes like this are gradually replacing even microwave towers as long-distance phone companies increase their use of geostationary communications satellites for telephone service.

build relay stations 10 to 40 miles apart (depending on the terrain and frequency band used). Each relay received the straight-line path of a signal before it went shooting off into space. It then aimed its own amplified version of that signal at the next receiving tower from a new direct line-of-sight location. You can still see these towers filled with microwave dishes as you drive across the country.

Satellite Links. While satellite forms of transmission are enormously expensive to build (and place in orbit), the distance and traffic load capacities they have make them ultimately very economical for long-distance traffic.

Direct Links to Your Phone. People now want phones where cables can't go. To meet this need, the telephone industry made use of radiotelephone connections direct from an exchange to a phone in a car or in a briefcase. Don't confuse this radiotelephone technology with the cordless phones you buy for your home. The cordless phone transmits by radio only as far as the base unit connected to your wall jack.

Microwave radio transmission now not only connects local exchanges and long-distance service, it is also used to broadcast directly to a phone in your car or briefcase.

These are just some of the companies now competing to be your primary long-distance telephone service.

Long-Line Leasing

The actual ownership and control of these various transmission paths might vary also. Your signals might travel through facilities and exchanges owned by several different telephone utility companies. These companies lease circuits by the many thousands at special bulk rates. Then they resell their use to customers like you at attractively low rates. These new services are also free now to build their own long-distance links. Their long-distance rates are usually less than those of a regulated utility like AT&T because they are not legally obligated to provide unprofitable services to out-of-the-way places. Long-distance rates now, however, are not as dramatically different as they once were.

In the new deregulated environment, AT&T doesn't have to plow any of its long-distance income into the subsidy of local utility operations. In that sense, regulated AT&T is on a par with its new long-distance competitors. Instead, the remaining regulation emphasizes a requirement for the giant utility to continue to provide reasonably priced service to every community. Unlike its competitors, AT&T can't pick and choose only the profitable routes.

LONG-DISTANCE SUPPLIERS

It's important to know several things about these firms. One is that all but AT&T are essentially unregulated as to scope of service although they must still file pricing information with the FCC before any changes can be made in their rates. Savings have to be estimated on the basis of careful scrutiny by the individual consumer.

Quality of Service

The technical quality of their lines is likely to be about the same as that of AT&T. Advertisements that proclaim vast differences should be taken with a grain of salt. All can be used to reach virtually any local exchange in the United States or Canada. But many of the long-distance networks cannot be reached directly (that is, without resorting to AT&T long distance) in many smaller communities. This makes them useful and economical only on a selective basis.

Customer Choices

You are allowed to select one primary long-distance provider to be accessed directly from your home phone. You call in to your service by dialing the familiar 1 + Area code + Number. You can access other carriers (if you've subscribed to their service) by dialing 1 + 0 before actually dialing the area code and the number you wish to reach.

You may be caught in the middle of a media blitz with other phone subscribers in communities across the country. You see and read a barrage of TV commercials and newspaper ads urging you to make your long-distance service choice now, or "one will be arbitrarily made for you." Actually, the choice is not as permanent as the ads make it appear. Most local companies will let you change service once during the first year at no cost if the one you choose now doesn't work out for you. After that, you can change long-distance services as often as you like, by paying a small fee to the local company. And, you are not limited to a single long-distance service. You can have as many as you want (and can afford).

POINTS TO COMPARE BETWEEN LONG-DISTANCE CARRIERS (including AT&T, Allnet, U.S. Sprint, MCI, Western Union)

- □ **Dialing Access.** Most homes can access any of the major carriers conveniently now. Not the big deal it once was.
- □ **Prices.** The gap is narrowing month by month. Call Consumer's Checkbook, a nonprofit research organization in Washington, for information about a current evaluation of your most recent long-distance bill. Use the toll-free number 800-441-8933. The service will cost you from $10 to $75 depending on your bill.
- □ **Volume Discounts.** Considerable variance in billing practices here.
- □ **Sign-up Bonuses.** Shop around to see if anything attractive is currently being given away to new customers.
- □ **Operator Assistance.** Some do; most don't. This can be useful if you dial a wrong number and want your bill corrected on the spot. All carriers do have special hot-line numbers you can dial if you have difficulty using the service or completing a call.
- □ **Travel Calling.** Some companies don't change their prices if you use their service away from your home phone; others do.
- □ **Voice Quality.** The gaps are rapidly narrowing to no important differences here.

Added-Value Services

Some of the companies only offer what is legally termed an "added-value" service for data transmission to and from computers. These firms specialize in relaying computer signals from modem to modem. Some of them have the built-in technical capacity to adjust transmission standards between the two computers in question. Where this is not the case, manual adjustments have to be made by the sender — a vexing process, to say the least. (Note the discussion of this issue in Chapter 15.) It is this feature that gives them their added value.

Questions to Ask

The home telephone consumer should shop around before jumping into a contract with any long-distance provider of voice or data services. The questions to ask are:

- □ Do I have the correct tone phone equipment to use your service correctly? Is my local exchange correctly equipped?
- □ Will I have to pay my local telephone utility to connect you if I want direct access? (Chances are you will if you have previously selected another long-distance provider.)
- □ Are your services in any way limited to certain hours of operation or low rates? (Usually so for lower rates.)
- □ Do you have operator services? How do I reach them if I need help? Are they full-time? Or only during "business hours"?
- □ What procedure do I use to report "technical problems"? Are there limitations as to the hours of operation of this sort of assistance?
- □ How do I go about getting a billing problem cleared up? (Be tough on this one. You probably don't want to have to pay for an incorrect call and then receive "credit" on next month's bill.)
- □ How do I go about using your service from other phones? Pay phones?
- □ Will I need to set parameters for my modem? What are they? (Be sure you can set your modem software to these parameters. Be especially careful to ascertain baud transmission rate.)
- □ Give me examples of the rates you charge to call places I call at the times I want to call.
- □ How long will these rates be guaranteed?

4 Special Equipment and Services

Time was when home telephone service meant local and long-distance calling and nothing more. But with modern electronic exchanges have come several optional services that many households have found well worth the small amounts they cost each month. (By the way, the utility-supplied services except "call waiting" require tone phones and service to work properly.)

Call Waiting

How many times have you been on the phone knowing that another person might be trying to reach you at that very moment? If only you could hear the second call trying to connect!

With the *call waiting* feature you get a signal that there's another incoming call, momentarily excuse yourself from the first caller in order to answer the call and then return to the previous conversation while keeping the new connection in place.

When you order up call waiting, the utility company hooks your line into a special hardware bank at the local exchange. The apparatus — which you never even see or touch — lets you hear, answer and hold a second call while remaining connected to a previous one. You hear the second call as a sort of distinctive sound on the line. You press the switch-hook (or disconnect button) quickly in order to engage the incoming call while leaving the first party waiting in silence.

When you have finished alerting the new caller to the fact that you are already on line with someone else, you depress the switch-hook again to return to the previous party. The second party can remain on "hold." As soon as you can complete your first conversation and the party hangs up, you are automatically reconnected to the second call again.

If you wish, you can use the switch-hook procedure to go back and forth between the two calls. And the system is designed so that the two callers cannot hear each other. Your conversation with each is perfectly private.

Call waiting is especially nice for families with teenagers who spend a lot of time on the phone with their friends. Then, when you call home with some important information, you have a better chance of getting past the busy signal without a long wait and many tries!

Call Forwarding

You have gone to a friend's home for dinner, knowing that an important call is likely to come while you are away. Your whole evening is strained by anxiety over the situation. You know of no way to alert your caller that you can be reached at a different number. That's a circumstance ready-made for the inexpensive *call forwarding* feature offered by many exchanges.

Call forwarding lets you route all incoming calls to the number you select. That important call would ring on your friend's phone (and the caller wouldn't even know the difference).

Using call forwarding is simple. You key in a couple of access code numbers, wait for the dial tone and then key in the number that will be temporarily accepting your calls. Deleting an instruction to forward the calls is just as easy.

There are a couple of cautions. In some exchanges, calls can be forwarded only to other numbers within the exchange. In most exchanges, however, calls can be forwarded to any number that can be direct dialed, including long-distance points. If you do forward calls to a long-distance point, your line will be billed for the cost of the call from your own telephone to the answering phone. The person calling you is billed only the cost of reaching your usual number (if, in fact, any billing is necessary).

Call forwarding is great for people who must keep in touch by phone. It's not a good idea for families that receive large numbers of routine calls. Imagine how embarrassing it would be if all of the "zillions" of calls intended for your teenager were automatically forwarded to a phone in a home where you were a guest for dinner! One of the weaknesses of the feature is that it cannot be "undone" from any phone but your own.

Three-Way Calling

How many times have you been involved in a discussion that needed a third party on the line? In the past, you could order a conference call from the phone company — but it took time and wasn't cheap. Now you can set up three-way conference calls from your own phone.

With *three-way calling* installed on your line at your local exchange, you are able to add a third person to a conversation by a simple process. You touch the switch-hook, listen for three beeps and dial tone, dial the desired number and then touch the switch-hook when the new party answers. Unlike call waiting, all parties on the line can hear and talk with each other. Disconnecting one of the three parties is equally simple. This is a wonderful feature when you need to get people together to talk.

Speed Calling

As fast as telephone systems have become, they can still be an irritation when you have lots of the same numbers to key in over and over again, day after day. The answer is an exchange feature that lets you key in anywhere from eight to thirty frequently called numbers (depending on the level of service you've ordered). You then can access these numbers with only one or two keys.

You might, for example, assign the number 3 to your best friend's home number. When you get ready to call the friend, you merely press an access key (usually the # on the key pad) and the digit 3. You can add, change or delete numbers at will from your own phone. You pay for the service on a flat fee, monthly basis.

You have to keep track of the speed dial number you have assigned numbers on your list. The phone company doesn't keep track of that information. A little directory list kept by the phone is a good idea.

Economy Rate Plans

Some telephone services available through a local exchange are in the form of "economy packages."

Pulse-Dial Service. In many utility areas, you may choose to order pulse service even though tone dialing is actually available. This makes good sense if you do little calling and no computer banking. Pulse service is cheaper per month than tone service. Pulse phones are usually less expensive to buy or lease. With some of the newer phones, you don't even lose the tone capacity for special services after you've reached your number with pulse dialing.

Flat-Rate and Measured Service. You may find telephone savings through the class of local service ordered up. For example, most utilities offer what they term a *flat rate*. This fixed monthly amount allows you to make as many local calls as you wish. On the other hand, if you don't use the phone much, ordering *measured* service may be wiser. Measured service is sometimes called "message units" or "message billing." Each of your local calls will be measured and charged according to call rates based on such factors as day of the week, hour of calling, length of the call as well as the distance to the called exchange.

A home-made directory is almost a necessity with speed dialing to remember the numbers you have programmed in.

Choosing a Special Rate Plan. Because these arrangements vary tremendously from utility to utility, you should check with your own telephone company to find out which plan is most economical for you. To save time, look at the information in the front of your local directory. There you'll find rates for the various savings plans in effect in your utility area. Otherwise, call the company business office and talk the matter over with a trained sales representative.

Automatic Answering

By attaching a special sort of tape recorder (called an answering machine) to your regular telephone, callers can leave a message when you're away from home (or unable to answer).

When you turn the machine on, it will automatically answer any incoming call by playing a short taped message, which you can record yourself. You might say, "Hello. You've reached 555-1234. I can't take your call at this moment but if you'll leave your name and number after the tone sounds, I'll try to get back to you as soon as possible." The caller will then hear a short tone burst, after which a spoken message can be recorded off the line onto your

An answering machine is a special kind of tape recorder that lets people leave messages for you when you're not home to answer the phone. The small device on top of the unit shown lets you retrieve your messages from any other telephone.

tape machine. When the caller hangs up, the tape machine turns off. When you return, you rewind the tape and listen to the messages that have been recorded, making a note of the names and numbers. It's like having an answering service — only cheaper and more private.

Answering machines range from the simple to the sophisticated. As do the prices. Some even have a special feature that lets you call into your own number and retrieve the recorded messages. This requires a special remote tone-generator (resembling a pocket radio) that sends control tones into your recorder, making it rewind and play out to your commands. One machine does not even require a tape to play back your prerecorded message. This one has a voice synthesizer, and after you select one of three basic messages, your callers are greeted with a synthesized voice that would say, "Hello. No one is available now. Please call after 7:30 P.M."

Amplified Telephones

There are two distinct reasons why some consumers might want an *amplified telephone* in their homes. For people with a hearing loss, it's often a matter of physical necessity. Regular telephone sound may be below the threshold of their hearing. An amplified unit can raise the sound level in the receiver by several hundred percent. These amplified telephones often have a switch on them so that other persons can use the same instrument without amplification.

There are other people with perfectly normal hearing who also want telephone amplification — but they want the sound to fill the room and not just to come through the earpiece more loudly. These people like to carry on a conversation without having to hold the telephone receiver. The speakerphones that amplify ordinary telephone sounds give you this freedom. You can talk while working in your shop, washing dishes, sewing or writing.

The same speakerphone units can also permit several people in a room to converse with a single caller — without having to pass the phone around from person to person.

The better speakerphone units are designed with specialized audio filters built in. These guard against feedback that happens when too much of the amplified sound from the speaker is picked up by the telephone microphone. The filters reduce the chance that the sound frequencies produced by the speaker will have a feedback effect on the pickup unit. (Feedback causes an annoying squeal that interrupts the phone conversation.)

Amplified telephones suitable for the hearing impaired are much more sophisticated (and expensive) than the low-cost speakerphones used by others. Speakerphones are commonly available from electronics stores while you must buy true amplified telephones from specialized telephone suppliers.

As technology becomes more sophisticated, more of these special features will be added to phones you can buy. For almost every communication need you have, there is now (or will be shortly) a telephone that will do exactly what you want.

Special telephones, or adapters for regular telephones, can amplify the sound level in the receiver for people with a hearing loss.

5
Your Computer's Link

One of the great moments in the history of computers occurred when it was realized that computers should communicate. This meant they could share programs and data with each other — and thousands of individual user terminals linked together by ordinary telephone lines.

Why Computers Talk to Each Other

The industry was quick to exploit the new possibilities of communicating with computers. For example, a terminal at the ticket counter of one airline at an airport could be used not only to book flights on that company's own large-scale, mainframe computer back at headquarters, it could also make reservations directly with the networked computers of the other airline companies across the country as well. The savings in time and labor — and customer satisfaction — were enormous.

A terminal at an airline ticket counter is connected by telephone to the computers of many different companies and travel agents.

But telecommunicating computers didn't stop with this single example. It was soon realized that travel agencies in every city and town could also be connected to the new booking network. Flight reservations in each airline's own computers could be made and confirmed from thousands of different locations. Even large airline users, such as giant corporations and government agencies, were sometimes allowed to bridge terminals into the system to expedite their own flight travel plans.

Then came the historic development of microcomputers, the smaller computers you use at home as well as at work or school. The early hackers (a term that had a more respectable meaning then than it has now) often had no other way to share programs to use on their computers except to tie their computers together through ordinary telephone lines, since the computer program industry had not yet recognized the market potential of this new communications and information tool.

Now, with increased sophistication in both equipment and use, it is possible for anyone to tie even the most modest computer into program exchanges, information sources and electronic mail banks all over the country — indeed throughout the world. You can now get to data banks and services set up by an industry like the airlines from your home. With special arrangements, as an individual computer owner you can dial into the reservation system through any one of a number of telephone-accessed data exchanges, using your personal credit card number as a billing address. Now, flight information and reservations can be almost instantaneously transacted with hundred of thousands — even millions of scattered locations! In many parts of the country, you can do your personal banking — even shopping — from the computer terminal in your home.

How Computers Talk to Each Other

You can link any two computers together through ordinary telephone lines with a device called a *modem*. This device converts the signals that computers understand to the kind of signal telephone lines understand.

Before the popularity of the personal computer, modems were typically used with standalone terminals. Terminals have a monitor

Personal portable computers can be used with any telephone, allowing instant data communication with home, office, data sources or even banking.

The mighty modem is used to allow any computer to communicate with other computers over ordinary telephone lines.

screen, keyboard and some connection to a large computer. Data are sent to the computer from the keyboard through a large bundle of wires, and then sent back and displayed on the built-in screen. By adding a modem to a terminal, someone could communicate with the large computer across town using just a pair of telephone wires instead of having a computer within reach of the multiconductor cables directly from the terminal. This is what the airline industry has done in the example we gave above.

Personal computers are not terminals. They were designed as general-purpose devices to be used in a variety of ways. Most personal computers today have a screen, keyboard and a connector called an RS-232C interface that allow them to act like a terminal and communicate with other computers through ordinary telephone lines.

Sending and Receiving

Communication can be defined as the sharing of meaning through commonly understood languages or *protocols,* using a common *code,* carried on some *channel.* This applies to both humans and computers. In human written communication, the alphabet is one code. The dots and dashes of Morse code are another. The English language may be thought of as a protocol. The medium of communication may range from the printed page to telegraph to radio. Some technology such as the printing press or radio transmitter prepares the code for transmission over the channel.

In computer terms, two or more machines are hooked together (the channel), sending each other information via a commonly understood character set (code) and format (protocol).

Just as there are many human languages, there are many different computer codes and protocols. These have been developed by different companies and groups for different purposes. Fortunately, microcomputers all use the ASCII character set (ASK-ee, standing for *A*merican *S*tandard *C*ode for *I*nformation *I*nterchange) and therefore understand a common code.

There are also three protocols that are now more or less standard for microcomputers. These are known as Bell 103A, Bell 212A and the newer CCITT V.22. The first two are standards developed by the Bell system that specify the timing and format for computer communication over phone lines. The third was developed for the same purpose by the Consultative Committee on International Telephone and Telegraph. There are many other protocols in the industry, however.

Communications Software

The only job a terminal is designed for is to communicate with a computer. Personal computers have to be programmed to do this. They

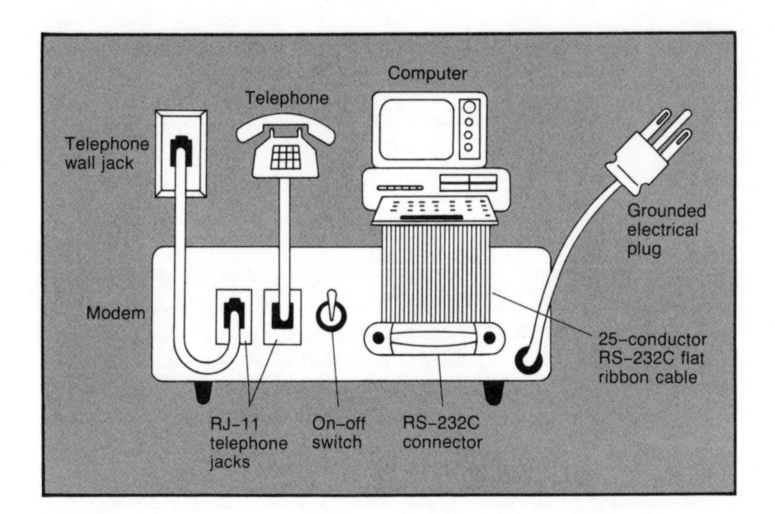

Typical connections on the back of a modem include a jack for the incoming telephone line, another jack for a telephone for voice communication over that same line, a large multiconductor cable going to the computer and the AC power cord.

have to be taught to act like terminals. When personal computers began using modems, programs were developed to make them emulate terminals. This software is typically called *communications software* and the basic purpose of one of these programs is to provide a *terminal mode.*

Coding Scheme Standards — A Common Language

In the best of circumstances, two computers can send and receive any language coding that they both use. The choices are wide: BASIC, FORTRAN, ASSEMBLY, COBOL — you name it. But both computers must be equipped to use exactly the same language coding. For all practical purposes, this is very likely to mean that the computers themselves must be the same types and models.

What happens if the computers are different, which is actually true in most cases? The computers involved can communicate only in text. Computer programs in languages like BASIC and FORTRAN cannot be easily used. This is really not as limiting as it may sound. Most computerists now don't engage in much programming of their own anyway. They are more concerned with messages than with process.

The trick is to have the sending computer convert all the text materials into the ASCII code — which, in fact, is likely to be the case to start with.

ASCII text is made up of letters, numbers and punctuation marks formed of bytes that have the same binary values for computer to computer, regardless of brand or model. Word processing, data base and spreadsheet programs customarily provide means for turning their outputs into ASCII formats.

Bits and Bytes. Integrated circuits in computers work by sensing voltage changes at the different electronic parts of the microchip. In binary terms a "1" bit is a high voltage (or "on") and a "0" bit is a low voltage (or "off"). It is this combination of ons and offs that gives electrical signals meaning in computers. In the ASCII system, each character (say an "A") is defined as a combination of seven on–off *bits* in the binary numbering system computers use. In bit code, "A" is 1000001, "B" is 1000010, etc. In the ASCII binary system, there are 128 different codes, ranging from 0000000 to 1111111.

ASCII coding is almost always given by a numeric value in the decimal, hexadecimal or octal number system. The ASCII bit arrangement for the character "A" is 65 in decimal, 41 in hexadecimal, and 81 in the octal system.

You may know that microcomputers work in eight-bit characters called *bytes*. The eighth bit can be used for various purposes, depending on the computer's architecture. In some it is used to tell the computer to display the character normally, or in reverse video on the screen.

Parity. The input/output device in the computer adds up the bits to be sent for a character, then puts in a *parity* bit to make the total come

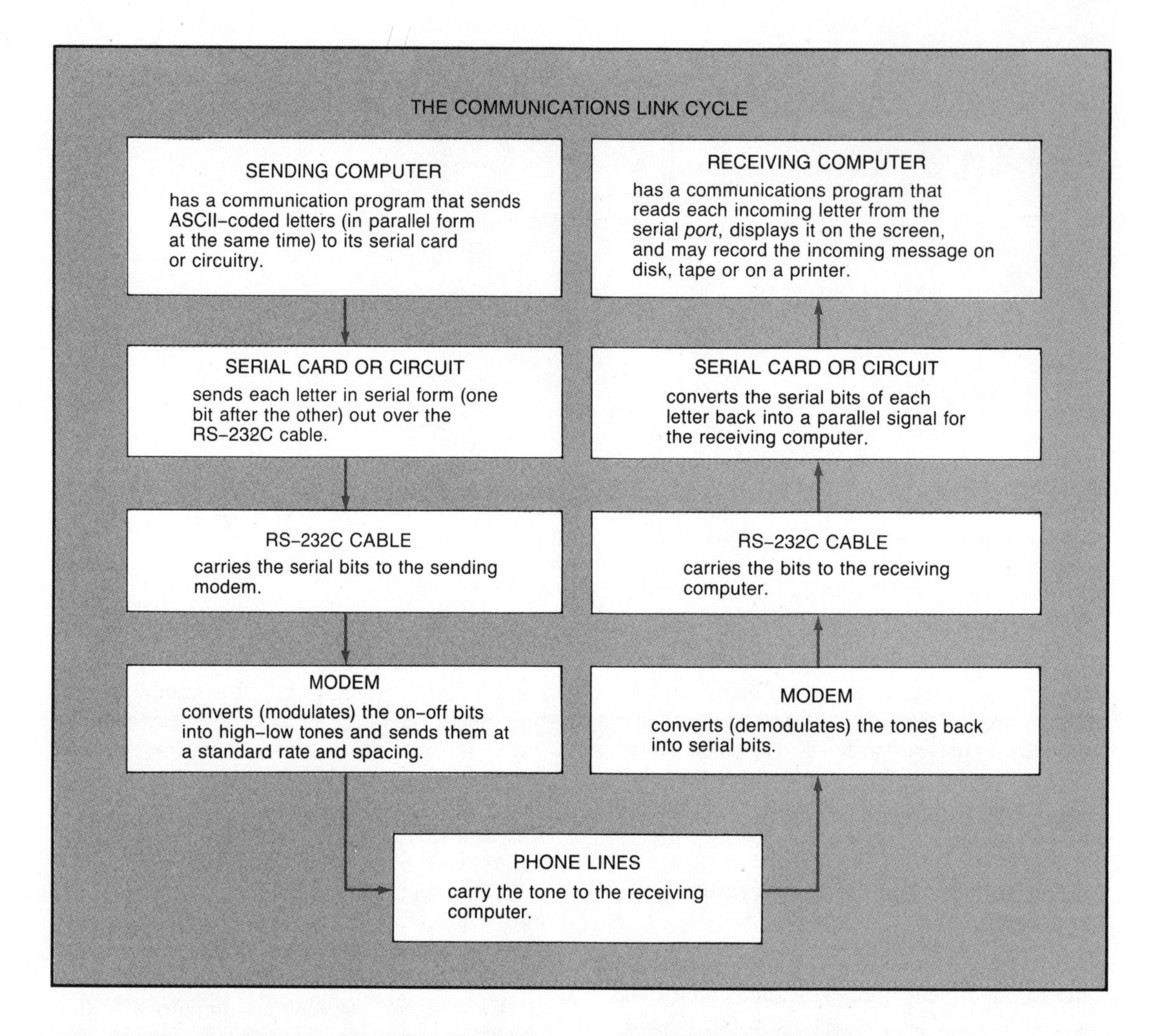

out either odd or even for error checking. Sometimes the parity bit is the eighth bit of a byte — sometimes it is an extra bit sent. If a receiving computer set for odd parity does not receive an odd number of bits for a character because some glitch in the telephone lines caused a bit to be lost, it will ask the sending computer to resend the signal for that character again. Parity error checking prevents a number 3, with a seven-bit ASCII binary code of 0110011, being mistakenly received as a 2, with an ASCII code of 0110010, because a bit was lost in transmission.

Start and Stop Bits. The modem will send out a signal saying it is ready to start the transmission of an ASCII character. This is called the *start* bit. There are also one or two *stop* bits added to tell the receiving system when all the bits have been sent and the code for one ASCII character is complete.

Serial Communication

The characters are sent *serially,* one on–off pulse or bit at a time. The architecture of most computers is designed to receive all eight bits at once, in parallel. If we send one bit at a time, as does Morse code with its dot–dash system, then we can use only one wire and the ground to send a signal. The printer on your microcomputer may use a parallel system with a separate wire for each of the bits in a flat "ribbon" cable. Both systems will have other wires used for control purposes and ground. It should be obvious that parallel systems can send signals faster than can serial systems since all eight bits go at the same time, rather than one after the other.

Error Checking and Duplex. There is also the issue of error checking. Often this is done by echoing the character received back to the computer that sent it. The term for this is full duplex

AMERICAN STANDARD CODE FOR INFORMATION INTERCHANGE
ASCII CODE

ASCII	Dec.	Hex	Binary	ASCII	Dec.	Hex	Binary	ASCII	Dec.	Hex	Binary
Space	32	20	00100000	@	64	40	01000000		96	60	01100000
!	33	21	00100001	A	65	41	01000001	a	97	61	01100001
"	34	22	00100010	B	66	42	01000010	b	98	62	01100010
#	35	23	00100011	C	67	43	01000011	c	99	63	01100011
$	36	24	00100100	D	68	44	01000100	d	100	64	01100100
%	37	25	00100101	E	69	45	01000101	e	101	65	01100101
&	38	26	00100110	F	70	46	01000110	f	102	66	01100110
'	39	27	00100111	G	71	47	01000111	g	103	67	01100111
(	40	28	00101000	H	72	48	01001000	h	104	68	01101000
)	41	29	00101001	I	73	49	01001001	i	105	69	01101001
*	42	2A	00101010	J	74	4A	01001010	j	106	6A	01101010
+	43	2B	00101011	K	75	4B	01001011	k	107	6B	01101011
,	44	2C	00101100	L	76	4C	01001100	l	108	6C	01101100
—	45	2D	00101101	M	77	4D	01001101	m	109	6D	01101101
.	46	2E	00101110	N	78	4E	01001110	n	110	6E	01101110
/	47	2F	00101111	O	79	4F	01001111	o	111	6F	01101111
0	48	30	00110000	P	80	50	01010000	p	112	70	01110000
1	49	31	00110001	Q	81	51	01010001	q	113	71	01110001
2	50	32	00110010	R	82	52	01010010	r	114	72	01110010
3	51	33	00110011	S	83	53	01010011	s	115	73	01110011
4	52	34	00110100	T	84	54	01010100	t	116	74	01110100
5	53	35	00110101	U	85	55	01010101	u	117	75	01110101
6	54	36	00110110	V	86	56	01010110	v	118	76	01110110
7	55	37	00110111	W	87	57	01010111	w	119	77	01110111
8	56	38	00111000	X	88	58	01011000	x	120	78	01111000
9	57	39	00111001	Y	89	59	01011001	y	121	79	01111001
:	58	3A	00111010	Z	90	5A	01011010	z	122	7A	01111010
;	59	3B	00111011	[	91	5B	01011011	{	123	7B	01111011
<	60	3C	00111100	\	92	5C	01011100	|	124	7C	01111100
=	61	3D	00111101	]	93	5D	01011101	}	125	7D	01111101
>	62	3E	00111110	^	94	5E	01011110	~	126	7E	01111110
?	63	3F	00111111	_	95	5F	01011111	DEL	127	7F	01111111

This is the American Code for Information Interchange (ASCII) established so computers of any make or operating system can share a common code of off–on binary signals to represent characters, numbers and symbols. The decimal equivalent shown for each ASCII character is the common base 10 number system, which uses the digits 0–9. Hex numbers are base 16, using the numbers 0–9 and letters A–F to "count" to 16 before increasing the digit to the left.

The binary system uses only 0 and 1 to represent a number or character. Each of the eight places represents a multiple of 2. To "count" binary numbers in the more familiar decimal numbers, add the place value of each digit that is a 1.

Decimal place values:	128	64	32	16	08	04	02	01	
The binary number:	0	1	1	1	0	1	1	0	
Decimal "count":	0 +	64 +	32 +	16 +	00 +	04 +	02 +	00	= 118 decimal

(with echo) or half duplex (no echo). If you are originating a message to another computer in the full-duplex mode, the characters you see on your screen are not being put there by your computer. They come from the computer you are talking to — even though that computer may be on the other side of the country and the signal goes through all the telephone systems between here and there and back. Most communication programs will allow your computer to display characters on the screen if you operate in half-duplex mode where the remote computer isn't sending back a signal for each character to your computer.

Protocol and Transmission Speeds

Transmission speed is expressed in *baud,* or the number of bits per second sent and received.

Serial transmission speeds over telephone lines are much slower than parallel speeds, where the two computers or computer components are physically wired together in parallel over a short distance. Most computer keyboards communicate with the central process-

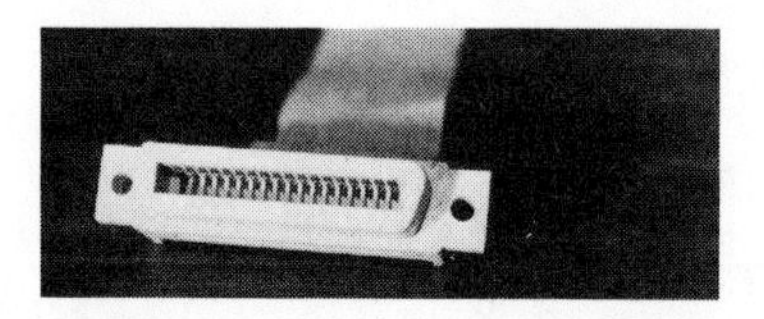

You can usually tell the difference between a modem cable and a printer cable by the kind of connector used. This type of connector is called a Centronics and is used to connect the computer to a parallel printer.

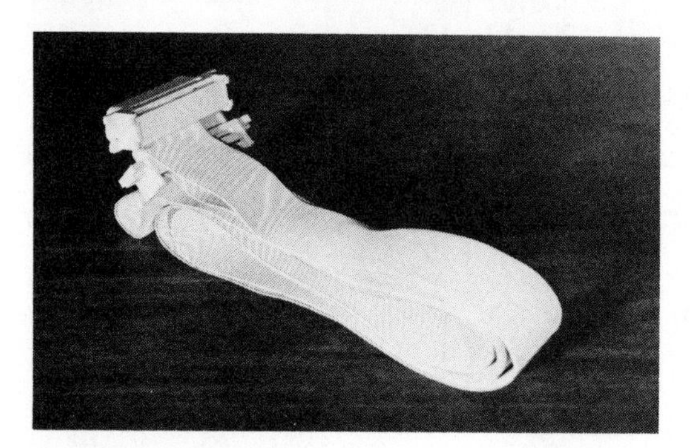

Ribbon cable used to send each bit of an ASCII signal in sequence to a modem for serial communication. Similar cable is used to send all eight bits at once for parallel communication, often used with printers.

ing unit (CPU) at speeds of 9600 bits per second (baud) or more. The simpler modems use a speed of 300 baud, under a convention established some years ago by the telephone companies. Newer systems will now permit speeds of 1200 and 2400 baud. While faster speeds than this are possible, very few modems for home computers are capable of using them.

The signal sent by a modem includes the start bit, seven or eight bits for the ASCII code, sometimes a parity bit and at least one stop bit. Under this protocol, as many as 12 bits can be sent for each ASCII character. At 300 baud, the computer can send out about 25–30 characters a second, depending on the number of parity and stop bits in the protocol. A 1200 baud modem can send out about 100–120; a 2400 baud will transmit about 200–240.

Bleeps and Bloops. In communications over telephone lines, the digital signals generated by a sending computer are converted into analog signals so they can be transmitted through linking systems to the receiving computer. There the incoming analog signals are reconverted into digital form again so the computer can deal with them.

Ordinary telephone circuits cannot transmit binary on–off signals directly. These conventional telecommunications links are designed to handle sound waves of varying intensity called *analog* signals. *Digital* signals are just patterns of high or low voltage in an on–off pattern. Even though most telephone networking between local exchanges and in the long-distance carriers is now via digital transmission, their equipment is set up to convert analog signals to digital before they send them out.

You recognize analog signals as tones; when the tones are combined and vary in rhythmic patterns, you hear them as music or voices. While analog signals are electronically more complex than binary ones, they are also much less accurate. This doesn't matter so much in the case of natural sound reproduction. The inevitable distortions in the analog wave forms, resulting from inaccurate transmission channels

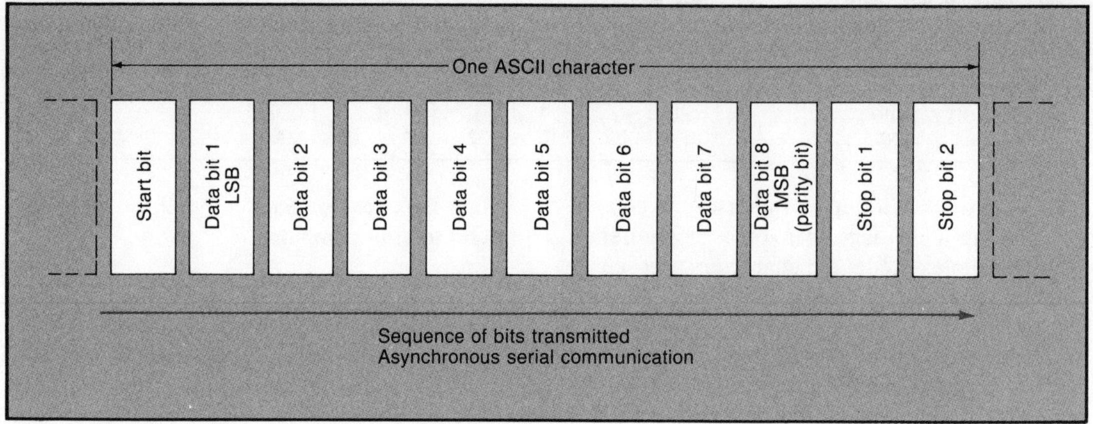

Your computer may have to send out as many as twelve bits for each seven-bit ASCII character code because serial communication requires adding information to signal when a character begins and when it ends, and sometimes to check if the right number of bits was received by the other computer.

and other factors, usually don't do enough damage to prevent our recognizing what was intended, even on low-fidelity telephone lines.

But one single distorted bit in a binary signal can change an "on" to an "off." That can alter the signal's meaning in a profound way. It can render a message meaningless to a computer that operates only on perfect binary signals.

Some device must modulate, or translate, the digital on–off signals from the computer into tones to be sent out over the telephone line. It then must demodulate or convert the tones back to digital form when it is on the receiving end. This *mod*ulator-*dem*odulator device is the *modem*.

Modems actually use two sets of tones, one for sending and the other for receiving (so the computer doesn't get confused and receive what it itself is sending). These are called the *originate* and *answer* frequencies. Most terminals "originate" when they send to a host computer, which "answers." If two microcomputers are hooked together, the operators (or the communications software program) have to decide who will use which set of tones. These tones give computer communications the distinctive sound of "bleeps" and "bloops" you can hear if you listen to a computer message on an analog telephone circuit.

The Bell 103A protocol specifies a speed of 300 baud and a specified range of frequencies for originate and answer tones. The Bell 212A protocol may also be used at 300 baud, but is almost always used only at 1200 baud and has a different range of sound frequencies. At 1200 baud, the originate frequency is about 1200 Hz and the answer frequency is about 2400 Hz. The newer 2400 baud modems for personal computers uses the CCITT V.22 protocol, which has still another range of frequencies for originate and answer. It doesn't take long to recognize the speed of transmission from another personal computer through your modem by listening to the pitch, or frequency, of the tones being sent.

Transmission Links

The telephone modular plug from a standard wall jack plugs directly into the modem. From your house, it connects your computer into the vast network of wires and other transmission links virtually anywhere on this globe — and beyond.

Computer communications, like routine telephone circuits, now use not only traditional twisted-pair copper wires but sophisticated coaxial cables, fiber optics, microwave beams and satellite relays. All of these telecommunication links do the same thing differently. They use electromagnetism or light to transmit a variable pattern of electronic energy through a carrying medium. It's the medium that makes them different.

Accuracy of transmission for computer communications involves the speed of delivery as well as the carrying capacity, or bandwidth, of the circuit being used. Think of the problem in terms of a liquid traveling through a pipe. The bigger the pipe, the more liquid that can pass through a certain length during a given interval of time. Conversely, the longer the interval of time the more liquid that can pass through the length of a given pipe size.

The broader the bandwidth of a telephone circuit (that is the more cycles it makes avail-

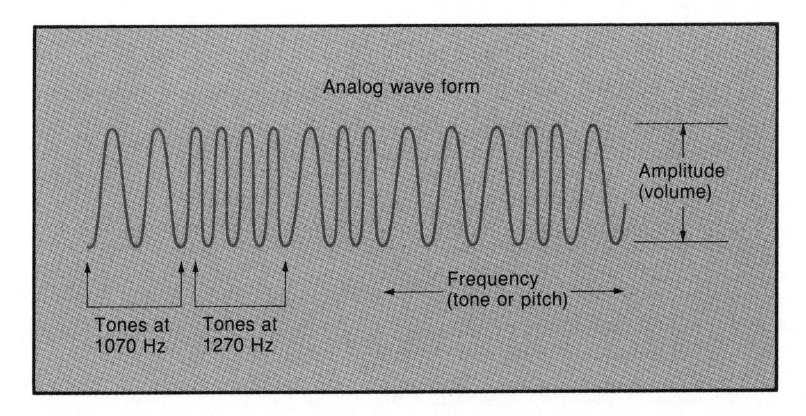

Tones and voices are sent over telephone lines as analog signals that alternately swing between positive and negative voltage, varying in both amplitude (loudness) and frequency (pitch) in proportion to the sound that produced them.

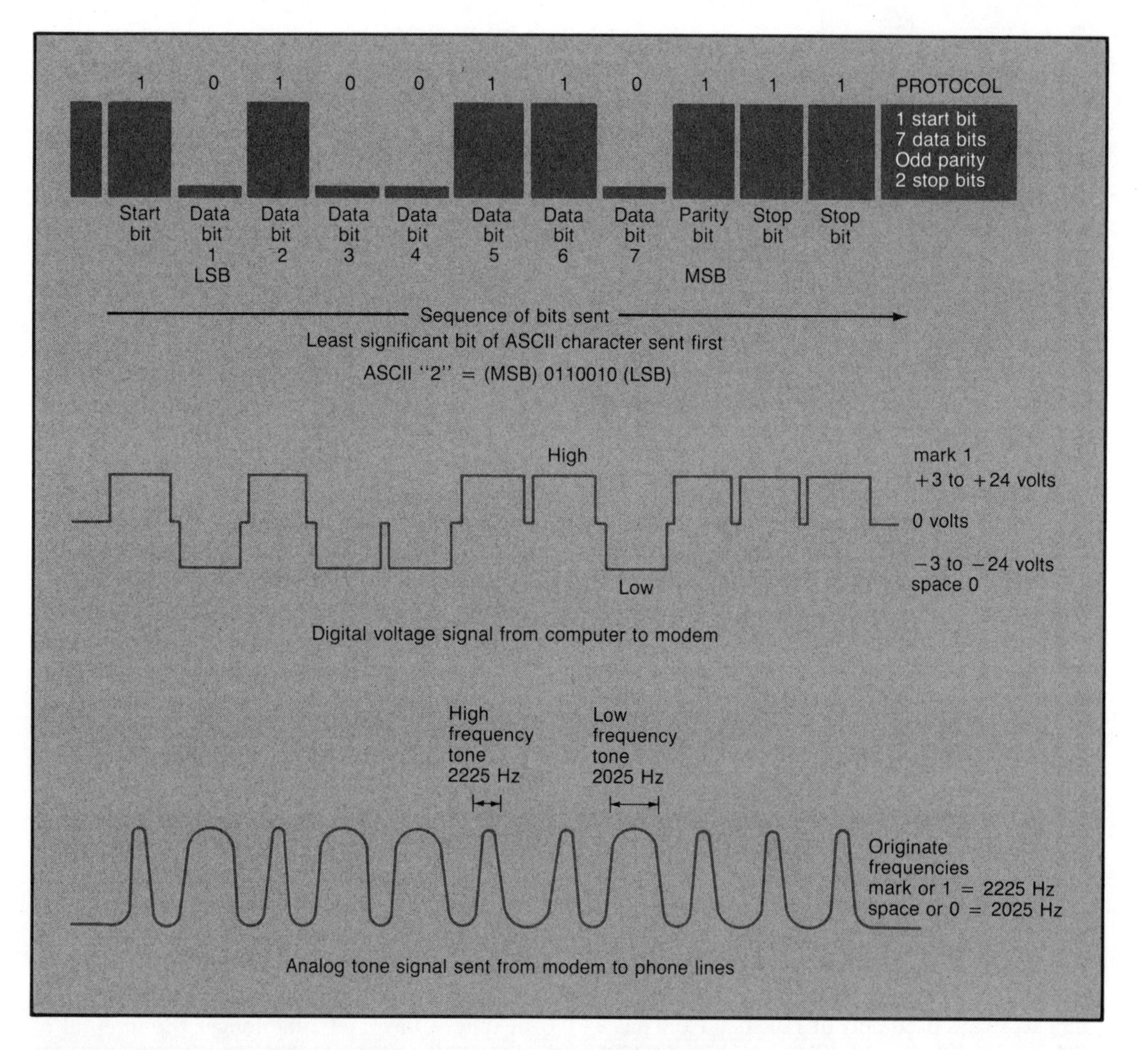

This diagram shows the relationship between the digital signals understood by a computer and the analog form of those signals that are sent out over the telephone lines by a modem. This example shows what happens after the key for a 2 is pressed.

All eight bits of the ASCII code for a 2 are sent at one time to a "holding box" integrated circuit in the computer (upper diagram). The serial board adds the appropriate start, stop and parity bits. On most computers, this generally takes only about 1/9600 second.

The serial board sends the bits for the character one at a time to the modem as a series of positive (high) or negative (low) voltages corresponding to each 1 or 0 of the ASCII code generated in the computer (middle diagram). This takes a little longer — 11/9600 second in this illustration — because the bits are sent out one at a time.

The signal is delivered to another "holding box" in the modem where each bit is translated into a tone (lower diagram). In this example, every high-voltage signal for a 1 bit is translated into a 2225 Hz tone and sent out by the modem over the phone lines. Every low-voltage signal for a 0 becomes a 2025 Hz tone. At 300 baud, it will take the modem 11/300 second to send out this one ASCII character code.

The computer finished its part of the job long ago (in computer time) and is just resting, waiting for the modem to finish and signal back through the serial board to send another character from the "holding box."

The protocol illustrated here is frequency modulation — a different pitch of the tone for each 1 and for each 0. Other protocols will use phase-shift modulation or even amplitude modulation, but the basic concept is the same.

able) the more of a message that can pass through in a given interval of time. Conversely, the longer the interval, the more of a message that can pass through a given circuit bandwidth.

Digital signals fed out from a computer travel at tremendously high speeds. The circuits they move through must therefore be wide in order to accommodate them at this speed. If the circuits lack the necessary bandwidth size, the signals must be slowed down to longer intervals. If this step were not taken, crippling signal inaccuracies would result. The relatively narrow bandwidth of conventional voice telephone channels (only about 4000 Hz) makes it necessary to slow down transmission speed through modems. At first, the fastest speed was 300 baud or less. As both telephone circuits and modem capabilities have improved, faster speeds of 1200 baud, 2400 baud up to 9600 baud are now possible.

Receiving the Data

Once the computer message reaches its destination, the receiving modem reverses things. It takes the analog frequencies from the telephone lines and translates them back into digital signals. After the modem checks them for accuracy, they are sent to the circuits fed by the RS-232C serial port, where they collect, one by one, until all eight bits of the standard ASCII code character are received. Then this byte is sent to your computer to process. It can then be displayed on the screen, sent to a floppy disk for storage or printed out on your printer.

Ways of Communicating

The communications software that accompanies a sophisticated modem will usually let the sending computer "download" or send materials to the receiving computer having compatible communications capacities. This means you can send a message — or a program — to another computer even if it is unattended. This is an essential arrangement if you are interested in sending or receiving *electronic mail*. Some modems now have a built-in buffer, or memory, that can collect electronic mail sent to you even when your computer is turned off.

Direct-Connect Communication. The basic process of connecting two computers involves calling the receiving source on the telephone and asking for the computer at that location to have its modem plugged in to the telephone line so the message can be sent, with or without keyboard replies. Long-distance communications can be a problem. It would take about 17½ minutes to send this chapter at 300 baud or 30 characters per second; just a little over 4½ minutes at 1200 baud. The faster modems are more expensive. Many computer users, however, actually find it cheaper to use the more sophisticated high-speed systems. That's because of the relatively high cost of conventional long-distance telephone lines. Over a period of weeks or months, the savings in long-distance charges for regular computer communications can often pay for the more elaborate modem — and then some.

Data Exchanges. A helpful variant on the direct-connect call-up procedure involves the use of what might be termed a *data exchange* as an intermediary. A data exchange is a mainframe computer used to store and relay messages and information to many users. Most such services are operated commercially. Computer user groups in many communities, however, have established similar services called *bulletin boards,* using large micros as the host computers. In the Washington, D.C., area alone, the list of bulletin board numbers makes up a

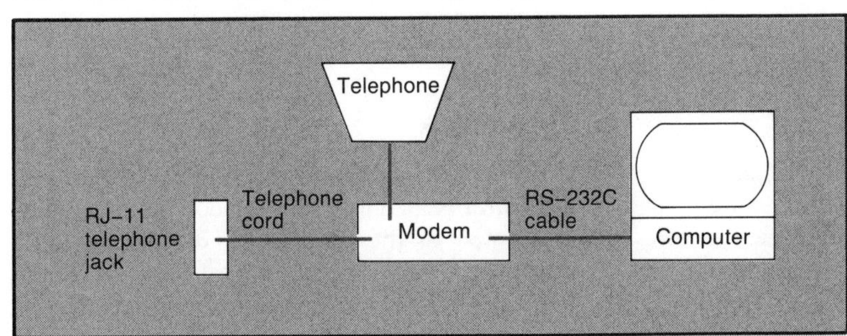

A modular cord connects the modem to the wall jack of your home telephone system. On some modems there is another RJ-11 modular jack for you to connect a telephone. When the modem is not in use, the phone can be used as you do any other extension.

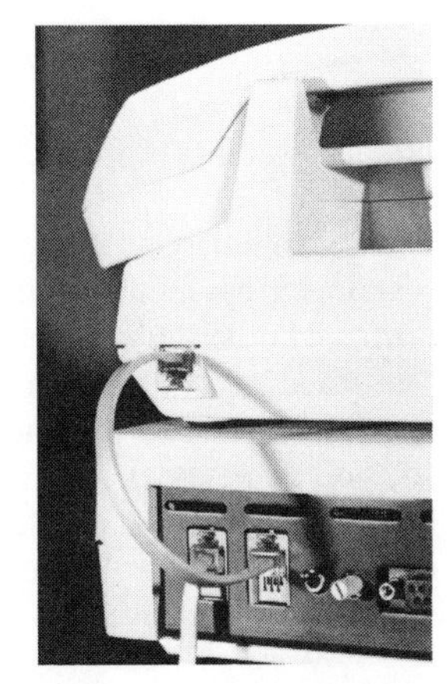

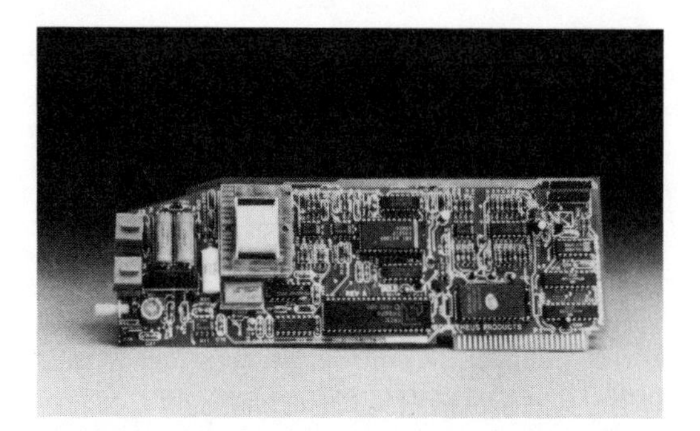

Modems are also made on an integrated-circuit card to fit the expansion slots of an IBM-PC, one of its compatible clones, or other computers like the Apple.

Computer communications over telephone circuits have become more reliable as phone companies switch over to new transmission technologies such as coaxial cable, microwave radio transmission, satellite relays and fiber-optic cable.

phone directory approaching that of a sizable village.

The commercial data exchanges earn their revenues from membership fees. Some are paid up front, others by use. Most of these modern "information utilities" make systematic use of special, lower-cost telephone circuits. But even so, you can expect a hefty charge for time online — unless you conduct your transactions during the middle of the night when the service's rates are lower.

Some modems have electronic storage spaces called *buffers,* where you can hold incoming messages until you have time to read them, or store outgoing messages until you're ready to send them out.

Security

Recently there has been a fair amount of concern over the fact that microcomputer owners — youngsters and adults — can use modem communications to break into mainframe computing centers in commerce, industry, government and the military. While the problem may have been a bit overdramatized in the news and Hollywood movies, it is perfectly true that such a thing can happen. It does take a great deal of skill (even if sadly misguided), patience and luck to accomplish, however. Furthermore, it's getting harder to do all the time.

Absolute computer security is difficult, if not impossible, to guarantee as long as the machine in question is linked to outside communications circuits. But relative and practical security is not. The use of various control numbers to limit access works pretty well if the numbers are complex — and random. Trying to have people use their birth dates, anniversaries or social security numbers as a computer combination invites logical deductions in tampering by the misguided gifted.

The Outlook

Most technological specialists believe that the computer is as important a telecommunications instrument as the telephone and radio. Its ability to send complex data in hard-copy, electronically sortable forms makes it a wholly different kind of information medium. Their confident prediction is that most people will use computer communications as often as they do the telephone within a decade or so.

6
The Choice Is Yours

Modern telephone instruments for the home come in a wide variety of styles, types and prices.

You can lease phones on a monthly basis from companies like AT&T as well as from commercial affiliates of many local telephone utilities. While the lease charges are not high, it is usually more economical in the long run to buy your own. (The best argument for leasing telephones for the home is that maintenance service is likely to be much more accessible — and often available without additional cost.)

Where to Buy Telephone Equipment

You can buy both new and reconditioned telephones from telephone firms like AT&T. Many of these companies operate phone stores at convenient shopping locations. The instruments they sell are likely to be of a very high quality. The range of styles is wide — all the way from POTs (telephone lingo for *p*lain *o*ld *t*elephones) to units cast in the shape of, say, Mickey Mouse or made to look like a French phone or the stand-up type from the 1920s.

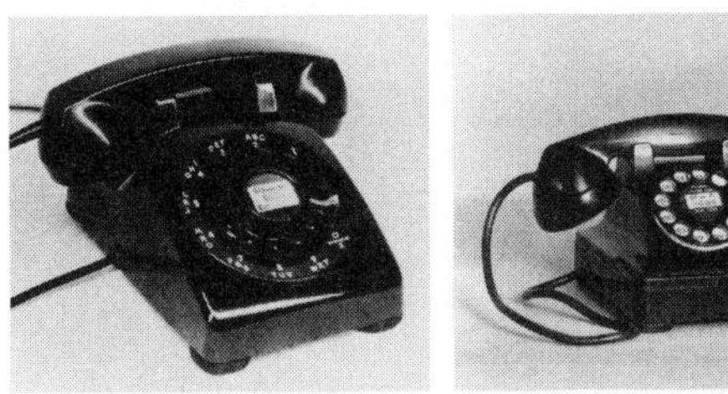

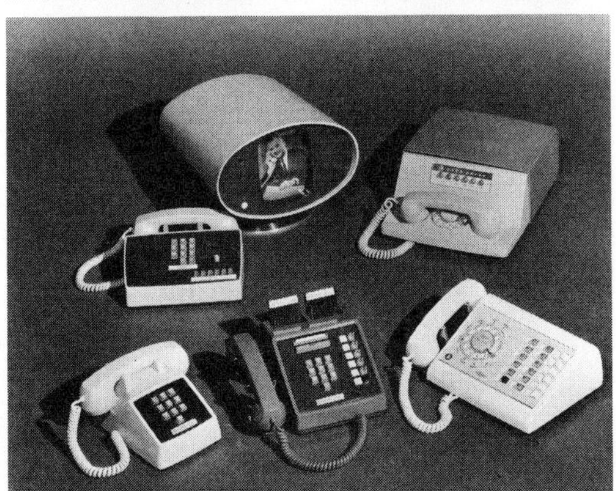

You can now buy telephones to match any decorating scheme or personal taste: phones that look like Mickey Mouse; phones that don't even look like phones; and even the old-fashioned basic plain black cradle phone.

There are specialty stores selling only telephones and telephone accessories, but you'll also find phone displays in electronic, hardware and department stores.

You can also buy telephones of all sorts from a broad range of sources other than the telephones companies or their affiliates. A list would include noncompany phone stores, department and discount stores, hardware and electronic stores, variety and drug stores. Even grocery chains and service stations sometimes get in the act!

The inventory of these stores usually includes both rotary-dial pulse and electronic-tone types. As a rule, however, the telephone company sales outlets offer phones only in the middle-to-higher price brackets. They rarely sell the "throwaway cheapies" which we'll discuss later.

What's Available?

Shopping for telephones is no longer an easy, decision-free chore. There are many styles you can choose from to match your own tastes and home decorating scheme. There are also features to match your communications needs and calling habits.

Kinds and Shapes. At the sake of oversimplifying, you have four basic shapes to choose from: First there is the familiar cradle phone with a handset that rests on a larger base where the dial or tone pad is located. These are descendants of the basic POT you know so well.

Cradle phones sit on a desk or table and have the familiar handset with the microphone and speaker separate from the base with its electronic parts.

Only now, you can get even this basic phone in a variety of exotic shapes and styles.

The second basic shape is patterned after the Trimline™ phone made by Western Electric for the Bell system in the days before deregulation. These are one-piece phones, which lets you set them down anywhere when you hang up. They work the same as the basic cradle-style phone, except they are much more compact and take up less room. They can also be harder to find under the mess of papers on a desk and are more prone to being left off the hook if not placed properly on a flat surface when you're done.

Wall phones make up the third general category of telephone types. These are units designed to be fastened to a wall mount instead of resting on a table or desk. Some inexpensive one-piece phones come with brackets that can be attached to the wall on which you mount the phone. These generally are not considered true wall phones.

All of the special-feature phones are grouped together as the fourth basic style. These include cordless phones that let you wander around the house — or patio — while only the base station remains tethered to the wall outlet. This category also includes the cellular phone you can use from your car.

Many new phones are patterned after the one-piece Trimline™ designed several years back by Western Electric. All the working parts are in the one piece.

Party Line Phones

If you have party line telephone service, you must have modifications made to the internal wiring of a phone before you can install and use it in your home system. These modifications

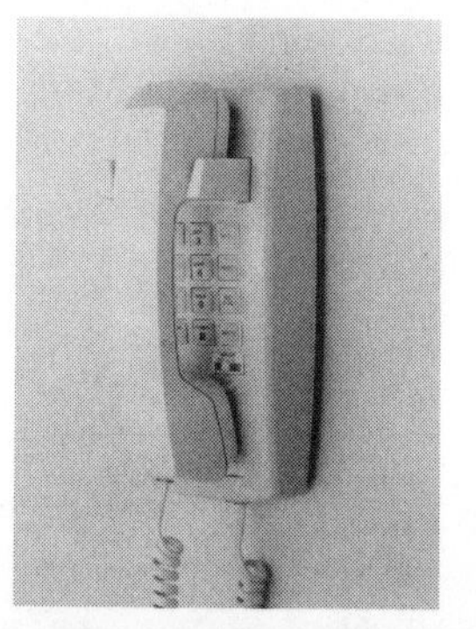

In many places in your home these wall-mounted phones are more convenient because they don't take up desk or counter space.

There are phones without cords you can use on the patio, at poolside or while working on your garden. They use a base unit connected to the phone lines.

will be (or should be) done by the store that sells you the phone; they are not wiring changes you want to make yourself.

On a party line, the local exchange "codes" the ring signal it sends out so that only the called number rings, without disturbing the rest of the phones on the party line. In some areas, too many phones are on the party line to allow discrete ring signals to each individual phone. In these rare cases, you may be assigned to one long and two short rings, while another home on the party line will be assigned two long and one short. You'll hear both rings, and have to listen for your own coded ring.

When buying a phone for a home system on a party line, you must first call your local telephone company and ask what ring number your phone number is assigned to. On a two-phone party line this will be either 1 or 2; on a four-phone party line you can be assigned to ring number 1, 2, 3 or 4. Give this information to the store when you buy your phone. They will change the wiring inside the phone so that it rings only when the local exchange sends out the "coded signal" for that ring number.

Phone Features

When shopping for telephones, keep in mind that they are competitive in both price and the range of features they offer.

Sound Quality. You buy a phone to communicate with other people. It stands to reason that whatever you buy should reproduce your voice and the voice of the person you're calling with the greatest fidelity and clarity possible. Listen for the volume of the other person's voice. Is the phone going to be loud enough — but not too loud — for you to hear conversations easily? Listen, too, for the sound of your own voice feeding back into the earpiece. This sidetone is important to keep you from shouting into the phone or lowering your voice to such an extent that no one on the other end of the line can hear you.

Dial. Of course you should look to see if the phone has a rotary dial or an electronic-tone pad. Some phones with tone dialing have a membrane covering microswitches instead of individual buttons you push. If you are considering a phone with this feature, be sure you are going to be comfortable dialing this way. When you're looking at the dial, be sure you're aware of where it is on the phone. Many phones now

Be careful when selecting high-style designer phones that may not have the groups of three alphabetic characters on the keys. Many firms are now advertising phone numbers using words, not numerals.

have the dial pad on the handset, where many people find it gets in the way or is hard to use to dial a number.

Some newer designer phones do not have the familiar groups of three alphabetic characters on the keys, but just numbers. This is okay, except when you try to dial (615) CLI-MATE, as many businesses are now advertising their telephone numbers.

Tone-Pulse Switch. A feature to look for in a phone is the tone-pulse switch. This lets you select between the older pulse dialing, used with the rotary dials, and electronic-tone dialing. If you have not elected to pay the higher monthly charge for tone service, this is a feature any phone you buy should have.

Ringer Switch. A handy feature to look for is a switch that lets you control how loud the phone will ring — or even turn it completely off if you don't want to be disturbed. Some phones even let you switch between the older bell sound and the electronic warbles and chirps of modern electronic phones.

Automatic Redial. Another feature on many tone phones now is automatic redial. This lets you redial the last number you tried with the press of only one button — a handy feature if you're trying to call someone and keep getting a busy signal.

Automatic Dialing. Some phones now have automatic dialing, letting you program in advance anywhere from ten to one hundred num-

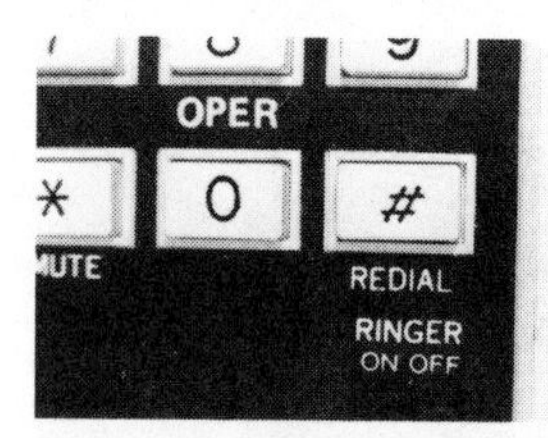

Some of the newer phones have this kind of ringer control, allowing you the choice of turning the ringer on or off, but not controlling how loudly it rings.

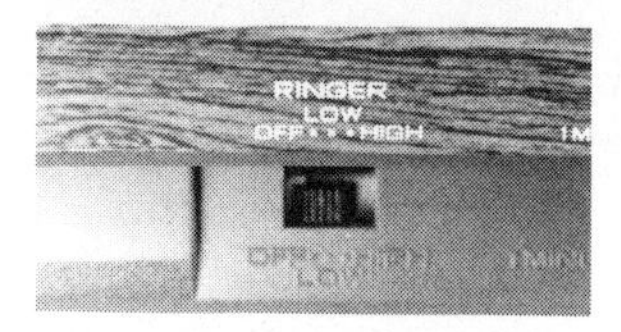

On many of the newer phones, you may find a switch that will let you select a loud or soft ring or turn the ringer off.

If your phone has the older clapper-type mechanical ringer, you can control how loudly it rings to some extent with this dial on the bottom of the phone.

bers (or more) you can dial at the press of one or two buttons. This feature is not likely to be found on the cheaper phones, but is readily available on phones in the moderate price range.

Hold and Mute. Some phones have a hold button that lets you hang up one extension and pick up the conversation on another while the calling

The *hold* key on a phone will keep your calling party on the line, but cut off the sound to them. You can use this feature to hang up one phone while you pick up a call on another extension.

party waits in silence. This mute feature also lets you "cut off" the calling party momentarily (without hanging up) while you say something to someone in the room with you. The electronic accuracy of this feature varies widely. It's best to try it out in the store first before you make a phone purchase decision based on this feature.

Flash Button. On many of the two-piece phones, there is a button on the handset that lets you hang up or disconnect from one call to place another without having to return to the base unit to engage the disconnect switch. This is especially useful for phones with very long cords and for people who like to wander when placing calls.

Digital Display. Some of the top-of-the-line phones will display electronically the number you're dialing, the time of day, and the length of time of the last call you made — handy if you're trying to keep your long-distance telephone budget under control.

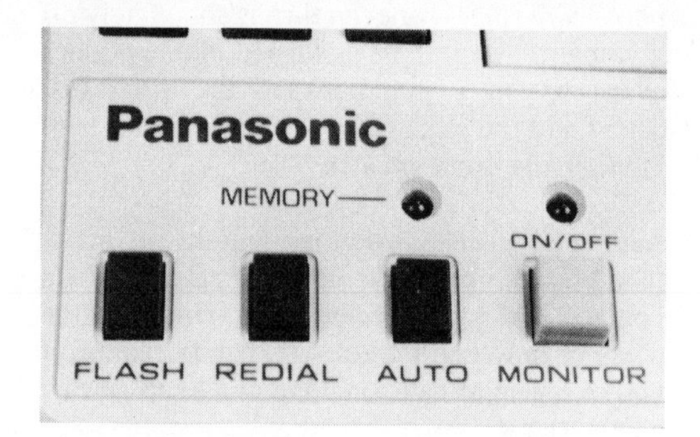

On some new phones you can automatically dial the last number you tried by pressing a *redial* key. On others, you can automatically dial numbers you have stored at the press of only one or two keys.

A more sophisticated phone feature will show you the number called, time of day or even how long your last phone call was.

For computer enthusiasts there are phones that have a modem built in.

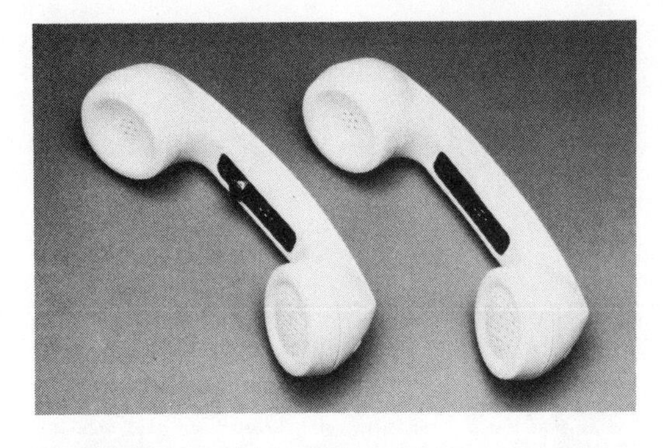

Phones with special amplified speakers are available for those with a hearing loss. The volume can be controlled by those who are not hearing-impaired.

Speaker phones let you conduct conversations while doing something else or when you want the whole family to be in on the phone call at the same time.

Phones like this are used when you have two phone lines (different directory listings) into your home. A key selects which line the phone is connected to.

Special Phones

Highly sophisticated telephones for very special uses are more likely to be found in some of the phone stores and through mail-order sources.

Modem Phones. Included among the many types of special phones are instruments that have a built-in modem for your computer. These may be phones that will accept two lines, allowing you voice communication on one and data communication on the other.

Speaker Phones. You can buy phones that have a built-in amplifier that lets you carry on a conversation without being tied down to the handset. The quality of sound from these varies widely, and checking for built-in features such as feedback suppression is a must when you try one out in the store.

Amplified Phones. Some phones come with an amplifier and volume control to control the level of sound for people with a hearing loss.

Multiple-Line Phones. There are phones that have connections for two lines. This is a feature becoming more popular in homes with teenagers. To use it, you must order two lines (different directory listings) from your phone company — and pay two monthly bills. It works like most business and office phones. When a call comes in on one line, that button lights up or flashes and you press it to receive the call on that line. These phones are more limited in styles and other features since they still are not sold as widely as the more traditional single-line home telephone.

Dial-less Phones. Right at the cutting edge of technology are the new voice recognition telephones that dial numbers when you tell them to. These phones recognize not only numbers as you speak them, but can remember the number for "Mom" or some other word you have programmed it to understand.

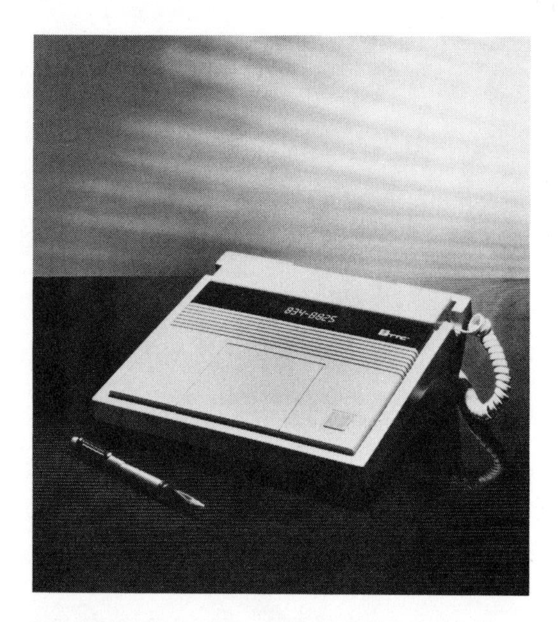

Some phones can dial a number just by your telling it whom you want to reach. *Voice recognition* phones are programmed in advance to recognize key words and translate them to a specific sequence of numbers.

Many discount stores, electronic outlets and even grocery chains sell inexpensive telephones. Most of these are cheap enough to throw away when they stop working and are not designed to be repaired.

During the programming stage, the voice recognition device analyzes your voice print, converts it to a digital code and memorizes the pattern. Any subsequent command is compared against the original prints. When the device finds one that matches, it executes a command. You can program the phone to dial 555-1212 in response to the word "Mom." There is an inherent security factor in this. The device can respond only to the voice patterns it has memorized. As an extra security measure, some of these phones require you to identify yourself with a code.

These phones are not completely dial-less, in the true sense. Most have concealed under a front panel a regular keypad you can use to dial numbers not programmed into memory and also to use when you enter new numbers. One model can store as many as 80 numbers.

Cordless and Cellular Phones. There is a whole range of telephones now that free you from the umbilicus of the lines. More help on the features of these cordless phones for your home or the cellular phones for your car or briefcase is in Chapter 8.

Repairs and Returns

Phone store customers are often told to return a broken unit to the store for repair. Customarily, charges are on the high side — once the warranty period expires. Service policies are sometimes offered to buyers at the time of purchase. These may be a bargain if your home is one in which the phone gets a lot of rough treatment from children or teenagers.

It is important to note that phone repair service — whenever it's offered by any seller — rarely means that a technician will actually come to your home to fix the problem. You will be expected to take the broken unit into a phone store or special repair shop. If the unit was purchased from a source other than a telephone company outlet, you will probably have to mail it off — at your own expense — to be repaired. Or you can sometimes do your own repair. (Chapter 17 has a full discussion of this topic.)

Buying TACs

Many of the places where you can now buy phones sell only very inexpensive phones, which can be accurately called "throwaway cheapies" (TACs). They cost only a few dollars, use keys to send out pulse dialing, have redial buttons, look attractive and come with a short preattached cord terminating in a standard modular plug.

TACs work perfectly well in many home situations. Most must be put down solidly flat on a hard surface to "unhook." This action causes a little bar on the bottom to press the hook closed.

Some come with a little slip-in wall bracket that lets you use the phone as a wall unit. You mount the bracket on the wall with screws (never nails, which are almost certain to pull loose over time!). Other TACs may proclaim the wall bracket is available as an optional accessory. Then try to find one that fits!

Phones that come equipped with a movable but substantial cradle base piece to force unhooking — like the Trimline™ phones from Western Electric — are nearly always better than TAC types. The same for real wall phone models.

All TAC styles are similar. The color is most often white; although, more and more decorator tints are appearing.

Quality of TACs

Their drawbacks are several. These TACs (usually imported from cheap-labor regions in the Orient) have noticeably inferior sound qualities. Some cannot be used for advanced telephone services which require electronic tones on the line. They wear out quickly. They break easily and, for all practical purposes, aren't worth repairing. You normally throw them away when they stop working properly. (Chapter 17 gives a few practical troubleshooting and repair hints.)

Because of their minimal construction, you can often hear irritating bloops and bleeps emanate from a TAC when a call is made from another phone in your home. Sometimes these "line echoes" are even loud enough to make someone else in your house think the TAC phone is actually trying to ring!

One of their more serious flaws is the placement of the switch-hook bar that disengages the phone line. As noted, most TAC units have been designed for solid placement on a flat, hard surface when not in use. The trouble is that as the phones gets older, the spring in the switch-hook begins to jam or stick. You put the TAC down after a call and walk away not realizing that the bar didn't push in far enough. The TAC is still connected into the exchange, tying up your line. That can become more than a nuisance. If it goes on long enough, the utility may have to disconnect your line temporarily. All the while, you may be wondering why nobody's calling you.

The same problem can even occur with a TAC used as a wall phone. If you don't force the unit into the plastic wall bracket firmly, you may not hang up properly. Some TACs have a red "in use" light on the back that lights when the line is open — or when the TAC is not hung up properly.

The switch-hook on these phones can stick, or the phone can be carelessly placed on some obstruction that keeps the line open.

All things considered — and assuming the TAC you buy has the required FCC certification and low ringer equivalence number (REN) — you may find the several dollars spent well invested. It's a nice cheap way to have a phone that works.

Where to Shop for Phones

You have as many choices now in your shopping trips for telephones as you do for your weekly groceries.

Shopping at Discount or Department Stores. Discount or department stores that set up a regular telephone sales display are likely to provide more than just TACs. In fact, they may inventory about as many "good" phones as a company phone store. And often at better prices.

These larger stores set up attractive displays of different types and brands of telephones and also offer a bevy of useful phone supplies for the home: extension cords, connectors, modular jacks and adapters. (As a rule, however, you may not be able to obtain longer lengths of real telephone four- or six-wire cable in these stores. That "raw" resource is more likely to be found in electronics, hardware or true phone stores.)

The problem with many of these "self-service" outlets is that if you need help from a knowledgeable clerk, you're probably in the wrong place. You may know a lot more about telephones than the salesperson does.

Shopping at Electronics and Hardware Stores. Electronics and hardware stores that feature telephone sales are likely to have the same inventory ranges as the self-service discounters. Their prices are almost certain to be higher, but their clerks are far more likely to be able to advise you on your purchase.

They will also help you select the right cords, connectors and plugs you may need to install the new phone exactly where you want it. As noted above, the variety of telephone supplies available in these and regular phone stores is apt to be greater than at other outlets. These are the spots to shop when you are getting into home telephone installations of real scale. Variable lengths of four- or six-wire cabling is more likely to be found in electronics, hardware or phone stores. The material in Chapter 7 will help you shop for just the connector or tap you're looking for.

Shopping at Phone Stores. Honest-to-goodness phone stores are probably the best places to shop if you really need expertly informed help before plunking down your cash. The disadvantage is that you may need more cash to deal with such outlets. They are apt to be "pricier" than other telephone sources. (Of course, there are fortunately always exceptions to this.)

Shopping by Mail Order. Buying telephones by mail (or TV) is possible. The risk you take is directly related to the reputation of the seller. As a rule, mail-order bargains are nothing more than TACs, closeouts on models a manufacturer has discontinued or unsold inventories of a manufacturer who has gone out of business. To their low costs, you must add the required mail costs. No real bargains there, most of the time. Try to avoid "clever" combination units like an alarm–radio–telephone whizbang. If the radio needs a trip to the shop, there goes the phone, and vice versa! Besides, if the phone component is nothing more than a TAC, the radio may last a lot longer, leaving you with a useless dead phone.

Shopping for TACs. When shopping for a TAC, pay more attention to the reputation of the merchant than the superficial appearance of the phone. If you get the TAC home and it fails to work, you want a quick exchange — or your few dollars back. You don't want a hassle with a slick huckster who insists you must return the phone to the "factory."

As to the TACs themselves, there's rarely a dime's worth of difference between them as to quality. Pay as little as you can — and to a store you can trust. Best advice: keep an eye peeled for the big Sunday newspaper ads from the reputable discounters. They often offer TACs at super-low "come-on" prices.

Tips on Buying

If you have decided to buy above the TAC level, here are a few points to keep in mind.

Sound Quality. Be sure the quality of the sound is acceptable to you. You buy a phone to communicate, not as a decorative home accessory. If you can't hear people when you call them, or they can't hear you, no styling considerations will make up for this serious deficiency.

Tone-Pulse. If you select a tone phone, be sure your local exchange offers electronic switching (nearly all now do), and that you have ordered tone service. Otherwise, it won't dial out at all.

Remember, keys do not necessarily indicate that the phone is a tone type. Many inexpensive key phones are rotary-dial pulse phones in disguise. And don't be fooled by sound. Many pulse phones add a tone-like signal that is only a cosmetic attempt to sound like authentic DTMF tone dialing. Read the label carefully.

Many models have a switch to select between pulse and tone. This is a necessity if you have pulse service from your phone company but want to tie into some of the new long-distance services requiring tone signals to finish the dialing sequence.

Pulse Phone Bargains. If you want a good pulse phone bargain, by the way, a new or reconditioned rotary-dial model from Western Electric, GTE or one of the other utility-related manufacturers may be the best choice. They work dependably, with decades of experience behind them.

Length of Dial Tones. Listen to the tones in the store as you dial a number on the keypad. Some phones emit only a very brief tone beep, no matter how long you press the key. This may be good enough to get you through the local telephone company's equipment, but frequently the long-distance services and some of the specialized services such as telephone banking or computer data sources require a longer tone to activate their switching equipment.

Ringing. The ringing should not be heard in the earpiece after you pick up the phone. In some cases lately, the ringing was so loud through the earpiece that it caused hearing damage. This is not a feature you're likely to be able to check in the store. Be sure to ask the clerk about it or read the literature that comes with the phone.

Construction. The heavier the phone, the better the construction is likely to be. (TACs are almost as light as feathers.)

FCC Certification. Be sure the FCC certification number and REN are visible. Also be certain to meet the law by supplying this information to your local telephone utility when you get your new phone home. Be aware of the ringer equivalence number (REN) before you buy. The lower the REN the more phones you can have on the same line in your home. Be wary of any phone with a REN of more than 2.0. Most are less than 1.0. The total value of the REN of the phones in your house should not add up to more than 5, so don't start your phone system with the handicap of one phone that has a very high REN rating.

Repairs and Warranty. Pin down the required repair procedure, both during and after the warranty period. And be certain there is a warranty period of at least 90 days.

Style. If you choose a "cute" or "high-style" phone, be sure you're prepared to live with it for a number of years. The more exotic the style, the higher the cost is apt to be. And the more difficult and expensive the repair job.

Cords and Connectors. The better phones have a detachable spiral cord between the hand-

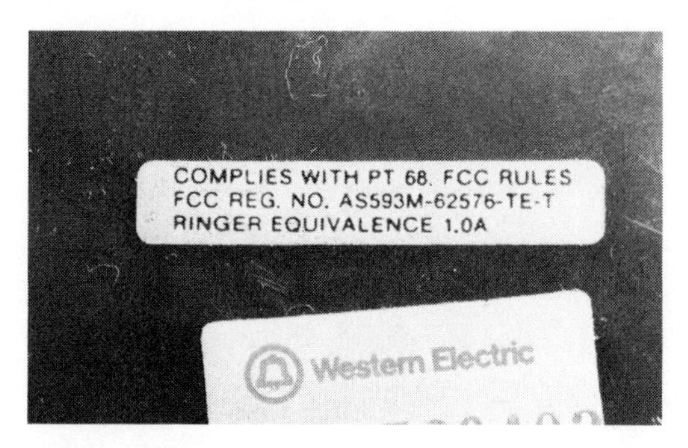

Don't buy any phone if it doesn't have the sticker that shows the ringer equivalence number (REN) and the certification that it complies with FCC standards.

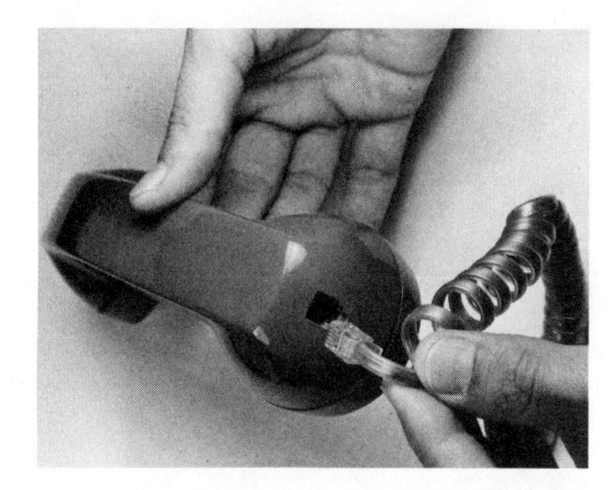

Better phones are completely modular with not only the modular cords to connect them to the wall jack, but modular spiral cords between the handset and base.

set and the base — as well as between the base and the wall connector. You can tell if it's detachable by looking for the little modular plug on either end of the cord.

If you buy one without a detachable cord, you may have to take the whole unit in for repair when the cord breaks inside because of hard or careless use. Replacing a detachable cord is simplicity itself. Replacing a nondetachable cord will require taking the whole phone apart. It's by no means the hardest job in the world but it will take a little time and patience. Moreover, if you don't pick a detachable cord model, you may not be able to use the phone as a basal connector for an inexpensive computer modem.

Cost. Let comparative costs guide you. The relationship between cost and quality is generally reliable in buying telephones. But always shop around for the best market price for a given unit.

Don't be overly concerned about where the unit was manufactured. TACs come from the Orient but so do lots of elegant instruments. And American-made units are not necessarily of superhigh quality either (although most are).

Take the reputation of the store carefully into account. Pick a store that will stand behind the merchandise — and won't hassle you if things go wrong.

Ease of Repair. Notice how the phone case opens, should the rare occasion ever arise when you need to make repairs inside. Be sure you have a screwdriver of the proper sort and size. Think about how easy the phone will be to keep

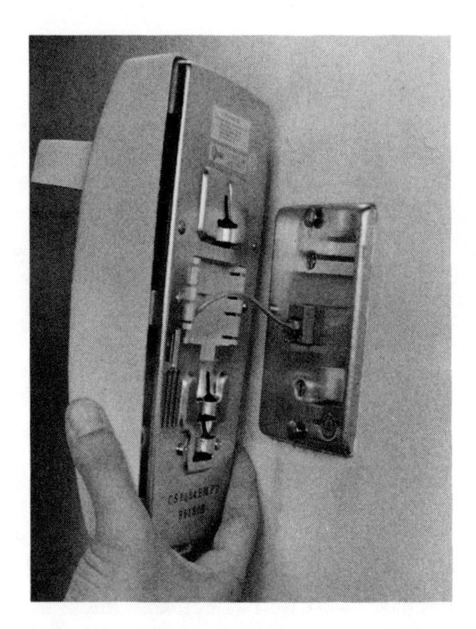

True wall phones require special mounting plates with large-headed, shouldered bolts that engage in the slots on the back of the phone.

clean. Fancy models with a lot of scrollwork can be a permanent and secure haven for room dust.

Wall Phone Mounting. If you're looking for a real wall phone (as contrasted with one that is placed in a wall-mounted bracket), consider how it is to be installed. Will you need a special modular connector plate on the wall immediately behind the unit itself? If so, can you get a line to the connector? (You'll find some help on mounting wall phones in Chapter 13.)

Try before You Buy. Ask for an actual "real use" demo of the phone if you can. Hear how it sounds; feel how it works under the fingers; sense how heavy it is. If this isn't possible, at least be sure you can return a higher priced model you simply don't like when you get it home.

If you're hunting for specialized telephone equipment like cordless phones and answering machines, look for tips in the other chapters of this book.

Phones like this have nostalgia value to the collector, but not much practical value without a ringer or a dial in today's modern home telephone system.

7
The Modular Connection

For many years, telephone utilities installed home telephones using permanent connecting cords between the instruments and terminal blocks mounted along baseboards. The manual installation procedure was so slow and expensive that the utilities lost money on most jobs, especially when it came to installing extension lines.

Consequently, the utilities developed a connector device that eliminated the need to touch the primitive terminal blocks. This new device was a plastic-covered box with four holes placed in a not-quite rectangular pattern on its outer face. Extension cords often came with four-prong jacks that could be quickly plugged into the wall connector. The time saving was an important economy for the utility companies.

As the courts began to move toward telephone deregulation in the 1970s, the utilities were forced to let customers hook noncompany phone equipment to their lines inside homes and businesses. Rather than let customers invade terminal blocks or even install their own four-prong extension cords, the Bell utilities devised a standard set of much simpler extension cords and connectors. The customer could get these new modular supplies cheaply for use with company or noncompany phones.

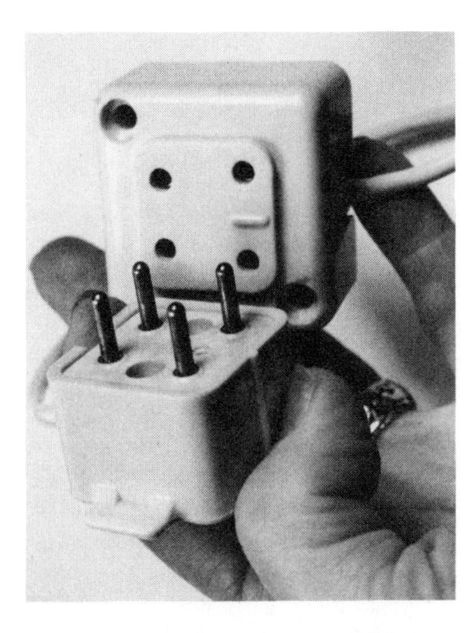

These four-prong jacks and plugs were used by phone companies when it first became possible to buy and install your own extension phones.

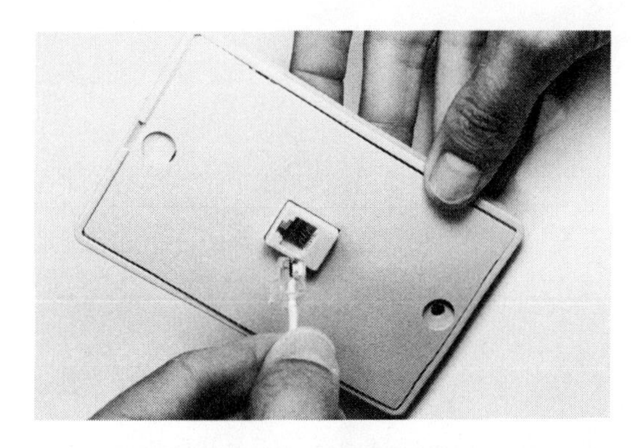

After complete deregulation of the phone industry, the modular connector system was adopted as a universal standard by phone companies and phone manufacturers.

The modular connector system makes it possible for you to hook phone equipment into wall outlets as easily as plugging electrical appliances into AC wall sockets.

SELECTING WIRING

The first thing you'll need is wires for your telephone system. There are two basic types to be concerned about. First, you'll need to select the right wiring for inside the walls or between outlets from room to room. Next, you'll need the correct wire between your telephone and the wall jack.

Telephone Cable

The wiring between outlets from room to room in your home is best done with color-coded cable available in rolls up to 100 feet long at most phone or electronic stores.

Four-Wire Cable. The most common cable for home systems is the four-wire cable. Inside a plastic sheath are four wires, each with different color-coded insulation: green, red, yellow and black.

Six-Wire Cable. If you have two telephone lines to your house (two directory listings), you will find it more economical (and neater installation) to use the color-coded six-wire telephone cable. A white and a blue wire have been added to the four basic color-coded wires. Use these where you would put the red and green on

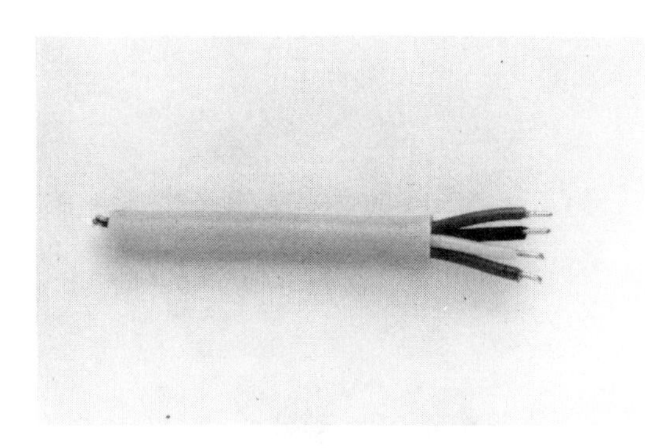

This familiar four-conductor, color-coded wire is used almost universally in home telephone system installations. Color-coded wires are essential.

Telephone cable with six color-coded wires is used in home installations where there are two different phone lines (directory listings).

the second phone connections, being sure the color coding is the same on both ends of the cable.

Two-Wire Cable. You may find some "telephone" cable at sale prices that may have only two conductors. While phones work on only two of the wires, the two-wire cable does not provide for adequate ground for your phone equipment and may give you (or the phone company) problems in the way your phone works.

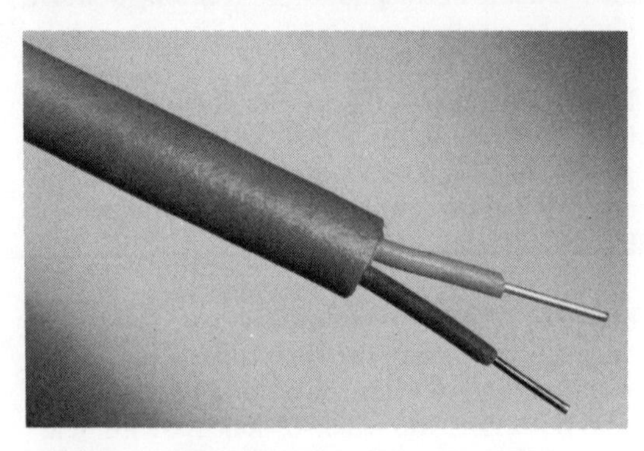

Special sale prices on telephone wire may be for only two-conductor cable that will not provide an adequate ground for your phone equipment.

Cords

There are three types of cords to select from as you shop for your phone supplies: telephone cords, extension cords and handset cords.

Telephone Cords. Telephone cords are designed to connect your phone instrument to a modular jack. A telephone cord will have a male plug at each end.

Extension Cords. Extension cords can extend the reach of telephone cords. Extensions will have a female jack on one end of the cord and a male plug on the other. The telephone cord from your phone plugs into the jack on the extension. True extension cords are hard to find. Most stores will sell you a long telephone cord (with plugs on each end) and an in-line adapter to connect them together. This arrangement

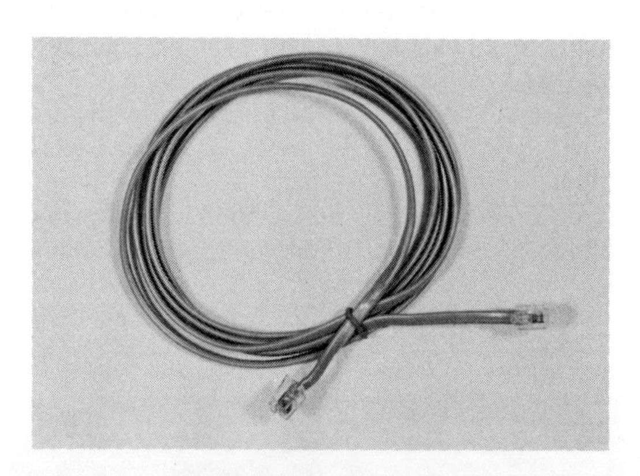

Telephone cords like this are used to connect your telephone to the wall jack. They come in standard lengths up to 25 feet.

A modular extension cord extends the reach of your wall jacks. These cords have a male modular plug on one end and one or two female jacks on the other.

works just as well, but it does give you one more place where trouble can happen.

Modem Cords. Many phone stores also sell very short telephone cords about 9 in. long. These are used to connect a modem to a telephone sitting on top of it.

Handset Cords. The cord that connects the handset of your phone to the phone base will most likely have smaller jacks than the standard RJ-11 type. Many people like to replace an older straight cord with a newer coiled version. You can buy handset cords that extend up to 25 feet from the phone.

SELECTING TAPS AND CONNECTORS

If you're embarking on an installation in a home where wiring was installed by the telephone company years ago, you probably will need to replace some of the basic taps inside your home. For new installations, you'll need the modular wall connectors to plug in your telephones.

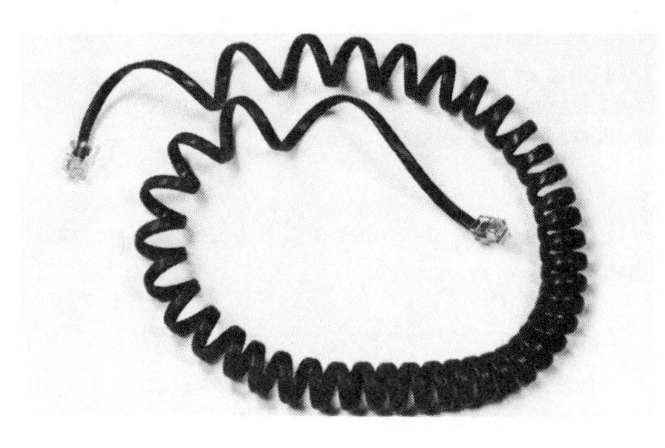

This cord connects your handset to the telephone base. Most phones come with a cord that can be stretched to 6 feet, but longer cords, up to 25 feet, are available.

This type of junction lets you branch out to five different locations in your home phone system using just modular cords.

The assortment of taps and connectors available can be confusing. Here's a catalog of the more popular types available with directions on where each can and should be used.

Wire Junctions

A wire junction lets you split the wires in your house to go to two different locations.

Modular Junctions. One type of junction has all modular connections, with a plug to connect it to a wall jack and four to six modular jacks to run cords to different phones or jacks in the house.

Screw Terminal Junctions. Another junction is designed for installations using bulk color-coded four-wire cable. Inside this junction are terminal screws for you to connect the wires that go to each of the two different locations. The input to these junctions can either be similar screw terminals or a modular plug.

Modular Jacks

Modular jacks contain the female connector into which the plugs on the ends of cords are connected.

Surface Mount Jacks. The most common modular jack is the small square box with the jack located on the edge or on the face. You use these where you run the wires along baseboards or walls.

Wall Plate Jacks. Where your wiring is inside the walls, you'll want to look for wall plate jacks. These are usually the same size and shape as the electrical outlet plates on your

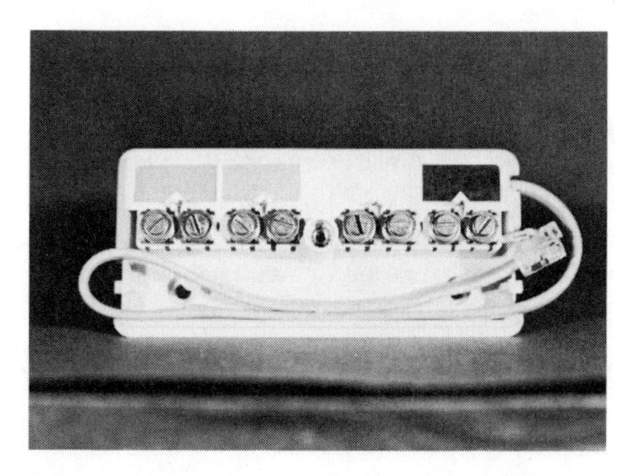

This kind of junction uses bulk, color-coded four-wire cable to run to other phone locations in your system. Screw terminals in the junction are also color coded.

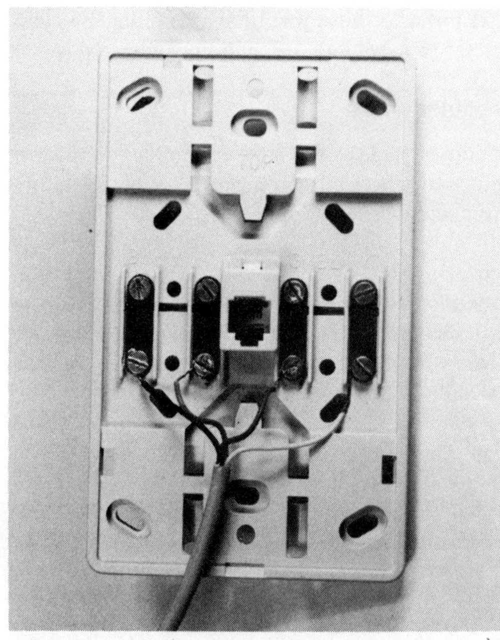

Modular jacks like this fit into holes cut in the walls, the way electrical outlets and switches do. Wires to these jacks are run inside the wall.

walls. They have one or two jacks into which you plug the telephone cord.

Wall Phone Jacks. Mounting units for wall phones resemble wall plate jacks, but they have additional bolts to fit the mounting hardware of wall phones. Usually you'll need a large-headed bolt with a shoulder that keeps it about $\frac{1}{8}$ in. away from the plate when tightened down. This

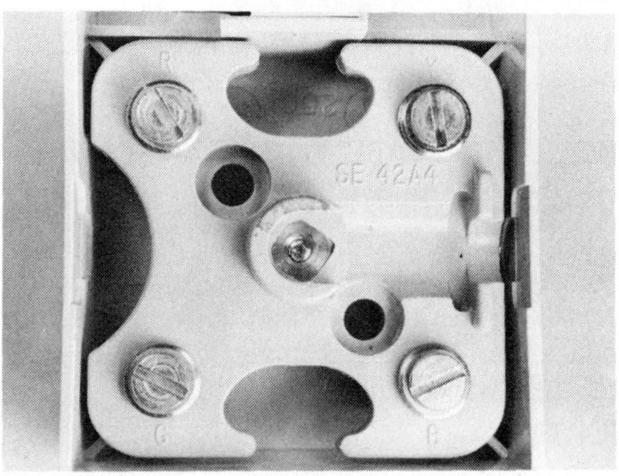

This kind of modular jack screws to the baseboard or paneling without having to cut into the wall. Wires are stapled to the baseboard or wall.

bolt hooks into a hole on the back of the wall phone.

Outdoor Jacks. There are also special weather-resistant phone jacks you can install outdoors. They have a flap that covers the jack when it's not used to connect a phone. For any outdoor phone installation it is highly recommended that you use these outdoor jacks to protect your system and the phone company's from any damage that water or weather will do.

Adapters

The modular system gives you a wide range of flexibility at modest cost. Where specific jacks, plugs or cords won't satisfy your phone installation requirements, you are almost sure to find an adapter that will help you do what you want.

Four-Prong Adapter. If you don't want to rewire an older four-prong surface mount or wall jack, you can buy a simple adapter to do the job. These plug into your four-prong jack and have a modular jack on the front or side for your phone.

Special mounting plates are used to hang wall phones. These have large-headed bolts with shoulders to accept the mounting slots on the back of the wall phone.

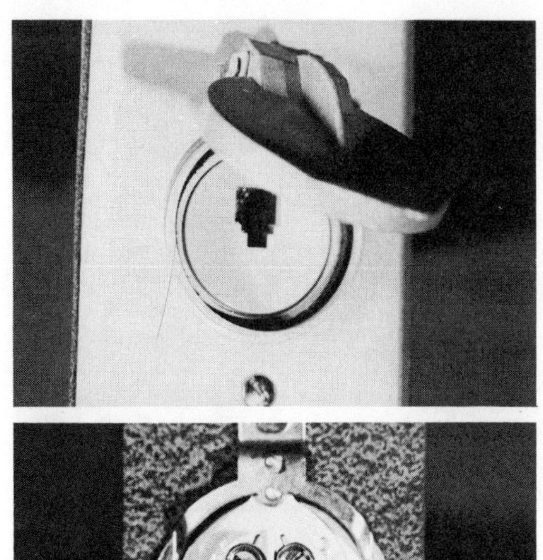

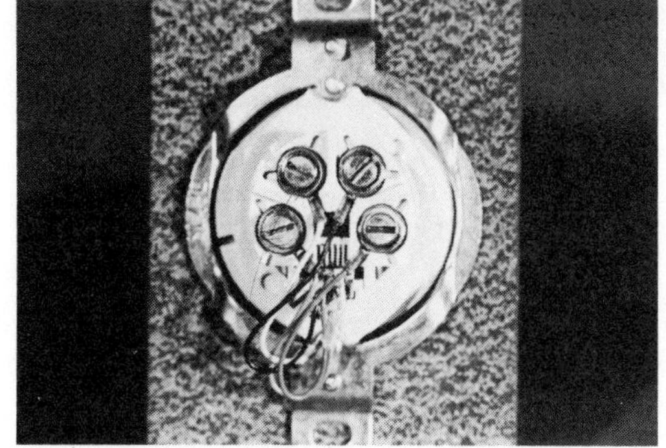

Special weather-resistant phone jacks are used for outdoor installations. These provide protection against the elements for your phone equipment.

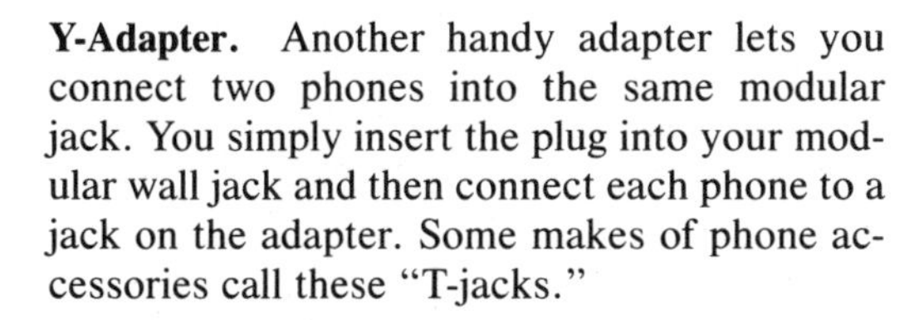

Y-Adapter. Another handy adapter lets you connect two phones into the same modular jack. You simply insert the plug into your modular wall jack and then connect each phone to a jack on the adapter. Some makes of phone accessories call these "T-jacks."

In-Line Cord Extenders. There is also a small unit with a female jack on both ends. You use these adapters to connect to the male plugs on two telephone cords to make your own longer extension.

Modular Converter. If you still have telephones without modular cords, you can buy a conversion plug to use it with your modular system. This converter has screw terminals to connect to the wires coming from your phone. The other side plugs into the newer modular jack.

A simple adapter like this will convert the older four-prong system to the newer modular without rewiring and installing new jacks.

TYPES OF CONNECTIONS

There are several methods used to connect the wires inside each of the jacks, junctions and some adapters. Be aware of this when you shop so you will buy the modular hardware with the kind of connections you feel most comfortable and competent to make. (You'll find help on how to make these connections in Chapter 12.)

Modular. Some hardware uses all modular plugs and jacks. You don't have to strip any

This Y-adapter lets you connect two phones into the same modular jack. These adapters can also be purchased as part of a modular extension cord.

These adapters have a female jack on both ends and are used to connect two modular cords together to make a longer extension.

wires, do any cutting or be concerned with any color coding to keep the wires straight.

Screw Terminals. Most of the connectors you'll buy have screw terminals to attach the new wiring to and modular connections to attach to the phone or rest of the system.

Knife-Type Connectors. Some telephone hardware comes with connectors that you use by just cutting the wire to the right length and slipping it into a slot made up of two sharp pieces of metal. The slot tapers toward the bottom and as you press the wire in, the sharp edges cut through only the insulation and make contact with the wire.

Snap Connectors. Many makers of modular phone accessories have made it easy for you to

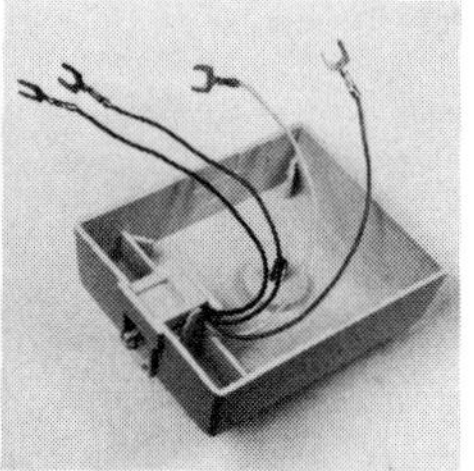

If you have an older system installed by the telephone company, you can buy adapters like these to convert these screw terminals to the modular system.

You can buy almost all hardware for installing your home system with modular plugs, jacks and cords if you don't want to bother connecting wires to screw terminals.

Bulk wire that you cut to length and connect to screw terminals on jacks and junctions is more economical to buy and results in a neater-looking wiring system.

connect to older-style terminal and junctions installed by the phone companies. These have snaps, very much like those you use to hook up to 9-volt batteries, that snap over the existing terminal screw.

TOOLS AND TESTERS

Most of the tools you need to install your modular telephone system are those you're likely to have around even a modest home workshop. There are some specialty tools, however, you may want to consider to make an extensive installation job quicker and easier.

Wire Strippers. Many of the phone stores and electronics outlets sell inexpensive wire-stripping tools. These have slots to strip the insulation off 24- and 26-gauge phone wire, a special cutter to remove the jacket insulation from four-wire telephone cable, a holder to hold small telephone staples as you nail them, and a tool to break out the tabs in some modular connector boxes.

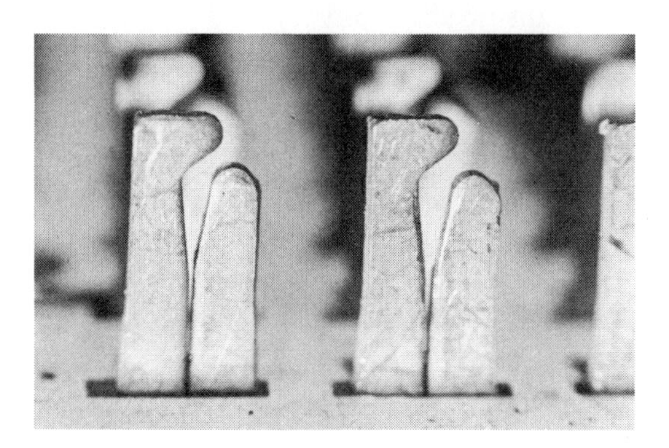

On this terminal, you press the wire in the slot and the knifelike sides cut through the insulation to make contact with the wire.

These connectors snap onto the terminal screws so you don't have to disturb any of the existing wiring on the jack or terminal strip.

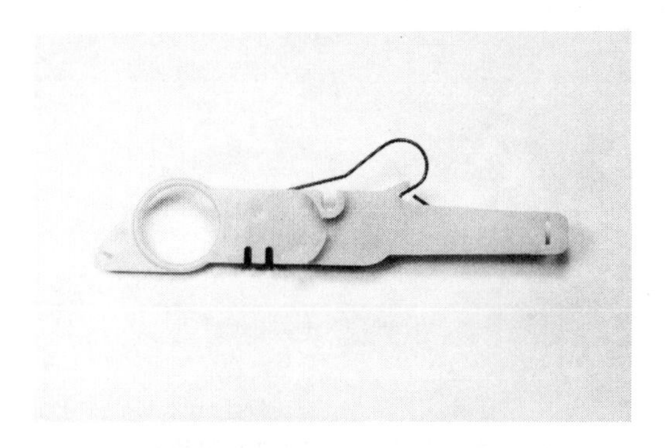

A wire-stripping tool is a handy device. The tool, with a built-in measuring scale, lets you strip a multiconductor cable without damaging the wires.

You can make your own modular cords and extensions to your exact measurement requirements with an inexpensive kit such as this.

Modular Plug Kit. If you want to try your hand at making your own custom-length modular cords and extensions, you can buy the plugs, cord and installing tool in a special kit at most phone and electronic stores. You simply cut the cord to length, then use a wire stripper blade on the tool to expose the four small wires in the cord. You insert the wire into the modular plug (following instructions carefully to keep the color coding straight) and squeeze the plug with the tool.

Line Tester. You may also find a unit called a telephone line tester. This is simply a lamp con-

This modular accessory lets you plug a tape recorder into your phone to make recordings of your calls.

nected to a modular plug. It lights up when plugged into a jack that is getting the proper voltage from the phone company. A working phone works just as well in testing if you have an extra, but this tester may be a good buy if you have only one phone.

MODULAR ACCESSORIES

At every phone, electronics, department or discount store you'll also find a variety of modular accessories you can buy for your phone system. Here is a sample of just a few.

Amplified Receiver. This replacement handset boosts the sound if your phone is located in an unusually noisy place or someone in your family has a hearing loss. There are also amplifier units that clip onto the phone or snap onto the earpiece.

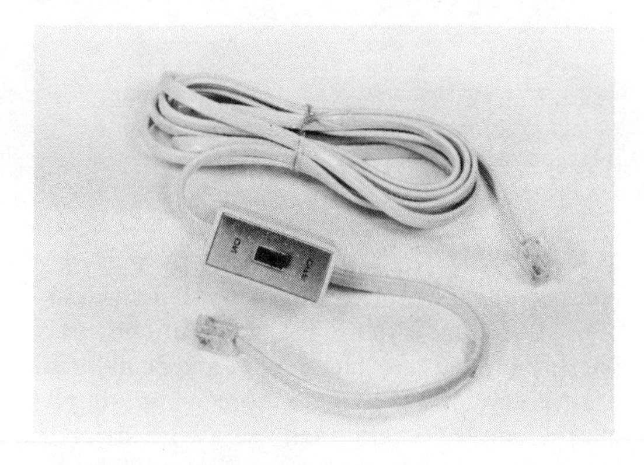

You can turn off the ringer on your phone with this accessory that just plugs into the line.

This extension ringer will let you know your phone is ringing when you're working or relaxing outdoors.

Recording Control. This adapter plugs into a Y-adapter with your phone on one end and into your tape recorder on the other, letting you make an audio tape of phone conversations.

Ring Silencers. If your phone does not have a ringer control, you can install this accessory to silence the ringer for privacy.

Ringer Extensions. There is also a modular plug-in unit that lets you extend the telephone ringer. This is handy if you're working in the garden or relaxing on the patio or poolside.

WHERE TO SHOP

Telephone stores and often many electronics and hardware stores are more likely to have the right cords, connectors and plugs you'll need to install your new phone exactly where you want it. These are the places to shop when you're getting into home installations of real scale.

Almost all connectors, and most wiring and cords, are packaged in "blister packs" that don't allow you to look too closely at the connections, or to try out an outlet with the plug from your phone. Some of the places where you buy your connectors may have samples broken open for you to be sure you have the right fit. Where this is not possible, and as a double check to be sure, follow the guidelines in this chapter to be assured of the right fit the first time.

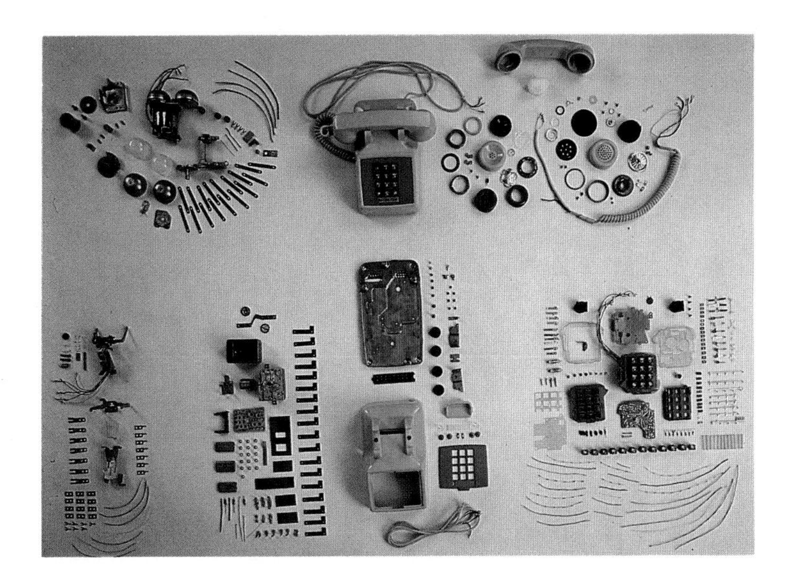

The inside of a modern telephone is a region where even the experienced professional treads with caution. Several hundred individual parts are assembled inside the deceptively simple device so familiar to us all. Many of these are delicate electronic circuits that are easily damaged by a misplaced tool or careless fingers.

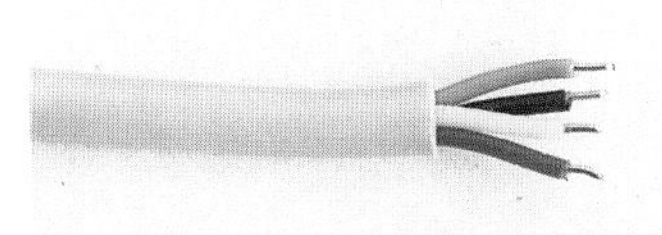

Four-wire, color-coded cable is now the universal standard for telephone system wiring. The cable has solid wires insulated with yellow, green, red and black plastic coverings. The yellow is a ground. Red and green wires carry the voice and ringing signals. The black wire is an extra, not normally used.

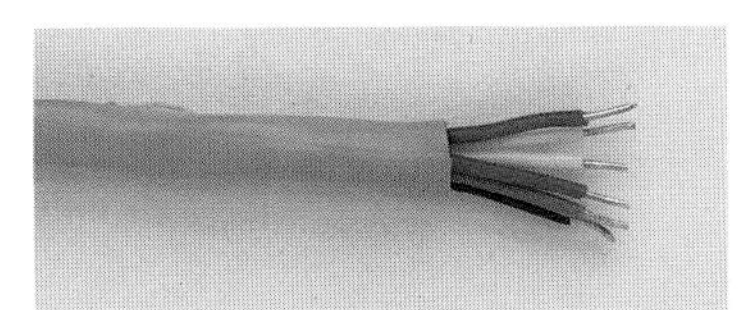

Six-wire cable is used in home installations where there are two phone lines (different directory listings) wired to the same locations. The blue and white wires are connected to the second phone the same way the red and green are on the first line. The one yellow wire is a common ground for both lines.

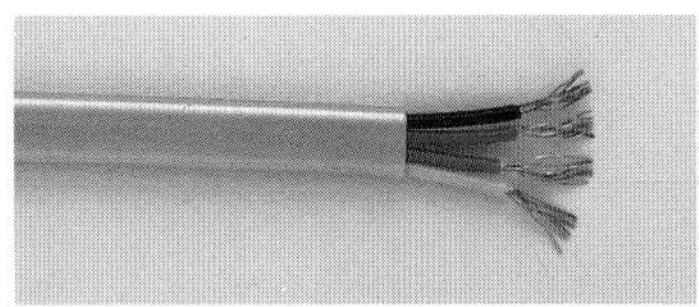

Stranded wire contains many individual strands of wire inside each color-coded insulated sheath. This gives the wire much more flexibility than solid-core wire. It is used in home installations where you need a cord that can bend and move freely without breaking. It is most frequently used to make modular cords.

Color-coded connections are essential in all home wiring. In every jack, junction or accessory you wire into your telephone system the red wire is always connected to the red terminal, green to green, yellow to yellow and the black wire to its companion black connection. Failure to follow this color coding can mean your phones won't work at all, they'll have inadequate ground or you can send signals or voltage out over the telephone company's lines causing potential damage to your neighbor's phones or the central office equipment.

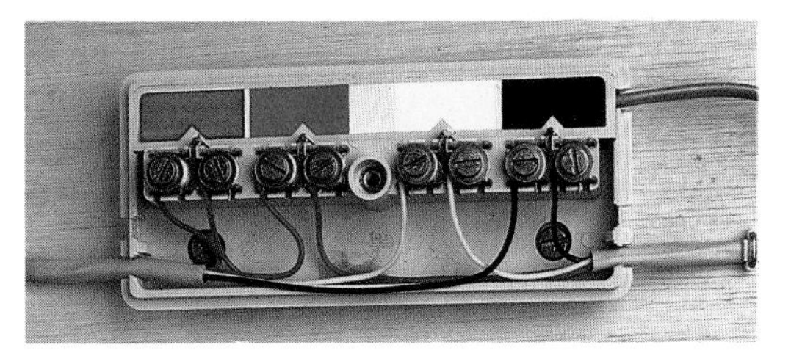

These junctions are used to split your system wire to several different locations. The modular junction (top) can split the incoming line to five different phones using just modular cords and extensions. Since modular plugs can only be inserted one way, proper color-coded connections are assured. The screw-type junction (bottom) is used with 4-conductor solid telephone cable to split an incoming line into two branches. The outbound green, red, yellow and black wires for each branch are connected to the terminal screw at the corresponding color-coded tab.

Typical installation of modular phone connections is shown on this wall model. A surface-mount jack (left) is appropriate in homes or apartments where in-the-wall wiring is not possible. Special jacks (center) have shouldered bolts to attach wall phones. Flush-mount modular jack (right) is ideal for locations where you can hide wires inside the walls. Jacks like these are available for almost any installation and often in colors to complement your decorating scheme.

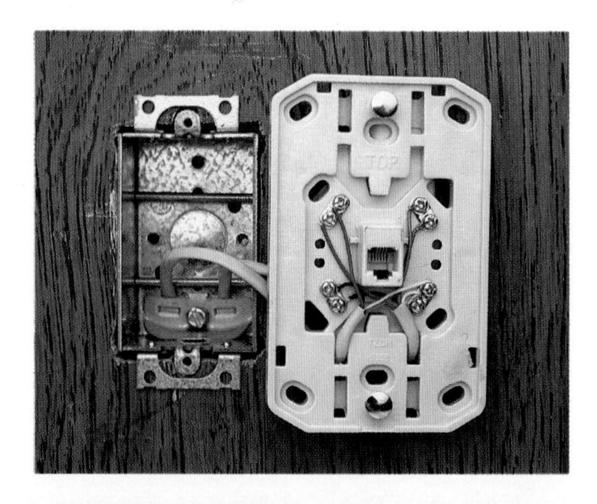

Wires to a surface-mount modular jack (photo above left) are stapled to baseboards, moldings or walls. Always use insulated staples. The kind shown hold the wire snugly but cannot damage cable if hammered in tightly.

Wall phone plates (photo above) will provide a more secure mounting if attached to a receptacle box anchored to the wall. Most modular jack wall plates come with bolts and holes to match screw tabs on these boxes.

Modular accessories of all types are available to extend and adapt your home system beyond basic wiring. Here a Y-adapter (photo at left) is used to connect a phone and modem to a single modular jack.

This model of a room wall shows typical wiring from the phone company's terminal at your house to the different types and locations of phone outlets. The network interface and wiring junction (right) split the system into two routes. Wiring is run up the wall stud to a receptacle box (center). Modular jacks can also be screwed directly to the wall surface (left) over a hole cut for the wire-connecting projection on the back of the jack.

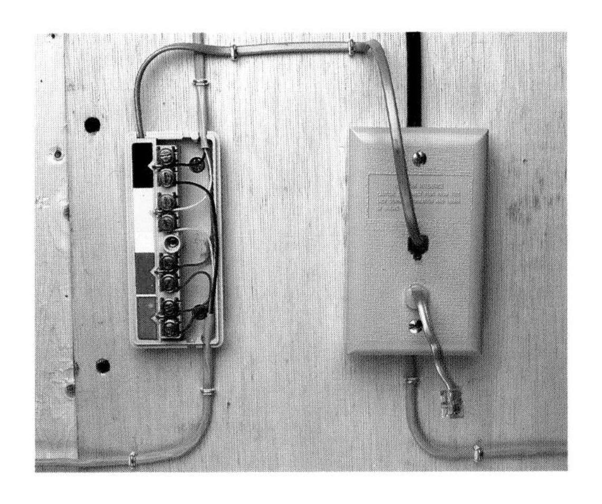

The line from the phone company (shown above) comes into the network interface. The junction has a modular plug to connect to the interface and screw terminals to which two different branches of your home wiring are attached.

A flush-mount modular jack (photo above right) has four color-coded terminal screws on the back to attach wires from the phone company's terminal or your junction. This jack is screwed directly to the wall covering over a large hole cut for the terminals.

Electrical receptacle boxes (photo at right) make mounting for modular jacks (especially wall-phone plates) more secure and give you easier access to the wiring for changes or checking. The receptacle box is nailed directly to the wall studs.

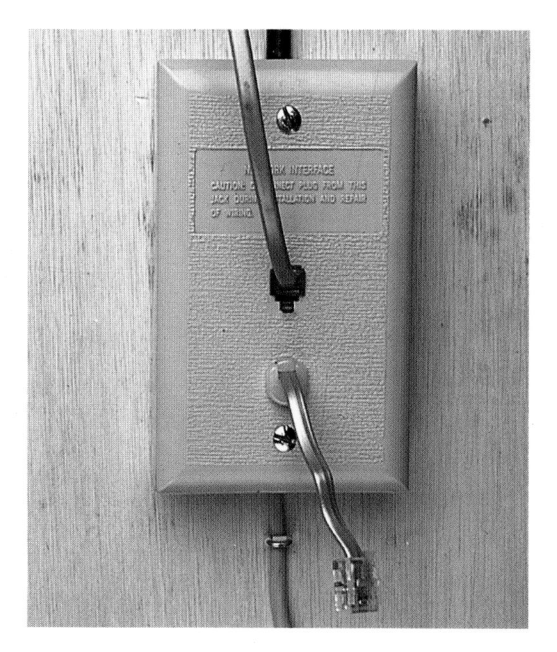

A network interface like this has a modular jack to let you disconnect your system from the phone company's lines when you are working on your wires or phones. The built-in jack (shown not connected) is connected inside the interface to a terminal for the wires seen coming out of the bottom. This allows either a modular plug (as shown) or screw terminal wires to be attached to the interface.

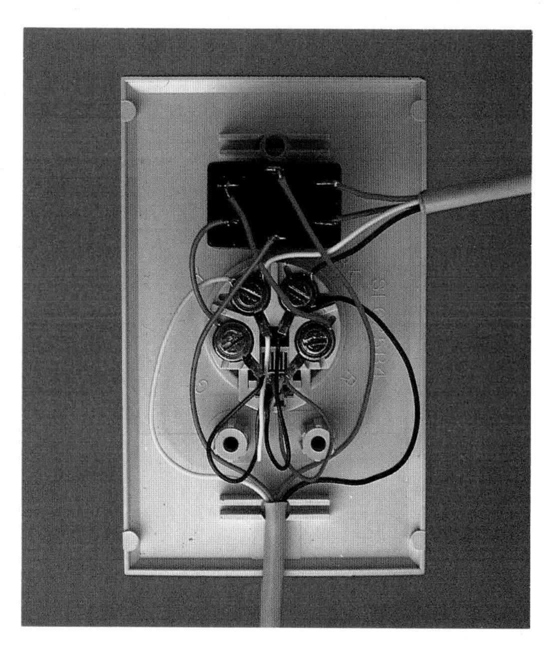

A switch like this mounted on a modular phone jack plate lets you disconnect other phones in your system for data security when you communicate with another computer over the phone lines using a modem. Your modem can, and probably will, confuse someone's voice signal for data if a phone is picked up while you're "on-line" and garbled transmission of your data will result.

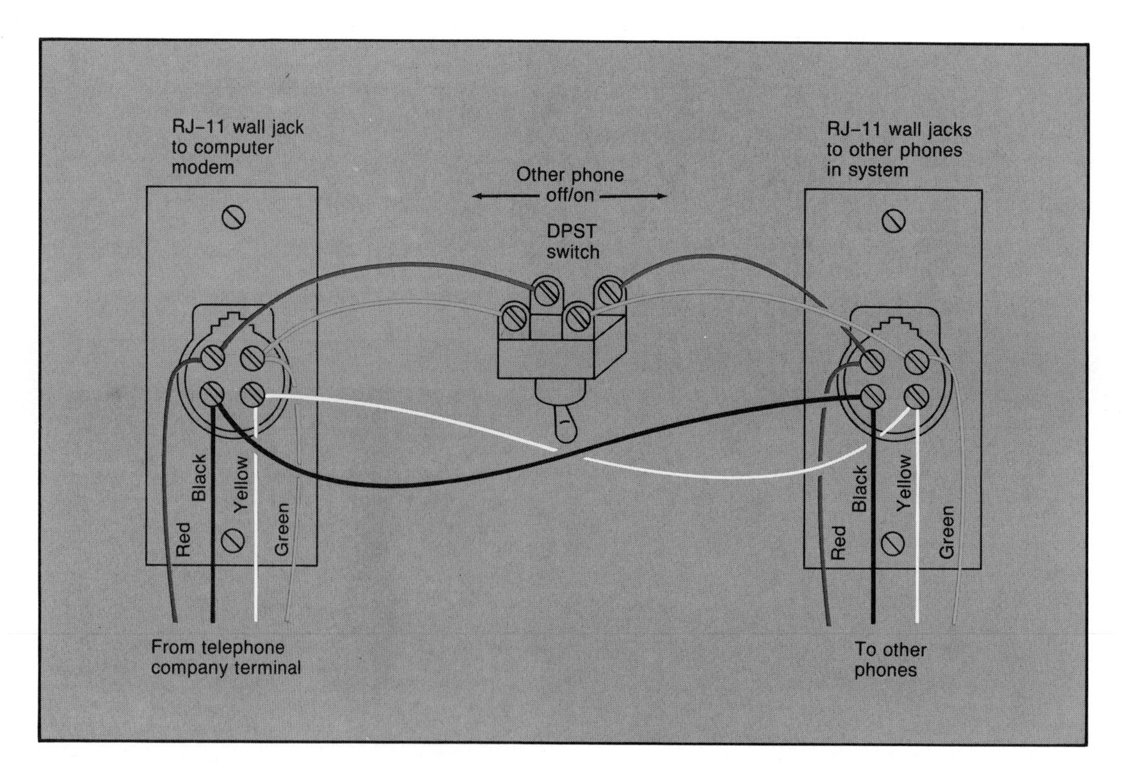

Correct color-coded connections for adding this data security switch to your system are shown in this drawing. The yellow ground and black wires run directly to each jack. The red and green wires go to the jacks through the switch.

When the switch is off, signals go only to the modem jack and not to other phones in your home. It is important to connect both red and green wires on the switch to the contacts with that same color-coded connection.

8
Phones That Don't Tie You Down

There might come a day when we'll wear telephones the way we wear wristwatches. That day isn't here yet, but miniaturization and portability has come to telephones. Cordless phones are now more reliable and conversations on them more secure. Cellular phones are the most portable. Some are still linked to the automobile but most are not and the umbilical cord is being stretched to its longest.

CORDLESS PHONES

The cordless telephone consists of a miniature radio *trans*mitter and re*ceiver* (a transceiver) built into a handset. The handset looks like one from any ordinary telephone with a tone pad, except the cordless one has a telescoping antenna. This hand-held device sends and receives sounds only the short distance between the transceiver and the base unit.

The maximum range between cordless and base station is about 1000 feet. The base unit is actually the telephone. It connects to the phone lines in your home the same way you would plug an ordinary telephone into a modular jack.

Early customers of the cordless phones had to cope with a lack of privacy in their conversations if their neighbors had similar units; they had to wait for the frequency channels to be free before they could place a call; and often they found themselves faced with enormous bills when unscrupulous freeloaders got within the range of their base unit and made calls on their phone.

Today cordless phones can be used almost anywhere. Calls go through even if neighbors are also using their cordless phones. The latest models will scan ten frequencies for a clear channel. Digital scrambling between the cordless receiver and its base unit in the home now ensures privacy. Moreover, no one with a similar cordless phone within range can charge calls on someone else's base unit without knowing the security number.

Cordless Improvements

Cordless technology is much improved today compared to what was available just a few years ago. This is due, in large part, to the switchover to the 46–49 MHz (megahertz) band from the 1.7 MHz frequency where cordless phones made their debut.

One problem that plagued the old 1.7 MHz phones was overcrowding. Only three send/receive channels were available on that frequency. If a lot of cordless phones were in use within range of each other, some users might not get a channel. Worse yet, they might get someone else's conversation.

The shift to 46–49 MHz eases the overcrowding by opening ten send/receive channels for cordless communication. Top-of-the-line units will scan the ten channels to find a clear one.

There's always the danger, though, that overcrowding in a tight vicinity might lead to inadvertent conference calls among neighbors. The best cordless phones use digital technology to scramble conversations at the base unit and unscramble them at the receiver end. And a security code must be used before an outside call can be made.

What's Next?

With cordless phones selling at the rate of 5 million a year, industry experts predict that overcrowding will be back to haunt us soon. Ordinary household wiring may help eliminate the problems of shared frequencies and overcrowding. Instead of using radio waves to reach the base station, some portable phones transmit conversations only the short distance to your ordinary household AC electrical wiring, from where it is sent to the base unit tied to the phone line.

CELLULAR PHONES

A cellular telephone call is transmitted to and from the dashboard of your car or from your briefcase via radio waves at a frequency of 800 MHz. The signal is picked up by an antenna at alternate cell sites, and relayed to the phone company's normal transmission channels. The easiest way to envision a cellular phone system is to imagine the honeycomb structure of a beehive, with the hive being a metropolitan area. Cell sizes vary, but as the caller travels from

The Mura phone shown has a multifeatured base unit. The cordless phone from Unitech has an AM/FM clock radio built into the base unit. Quasar's phone wakes you up to black-and-white TV.

The Comdial phone is a tabletop speaker phone that lets you wander as you talk.

one cell to the next, the signal is handed off from one antenna to another. The switching is done by a computer that monitors the strength of the radio signal to determine when to make the handoff.

When this concept was first conceived, a car was perceived to be the object moving between cells. These days it might just as easily be a person.

Cellular Phone Features

Though cellular phones are in more hands than ever, they are hardly a mass-market product. Prices for most basic car models are around $1000, while the phones with the most mobility are $3000 or more.

Standard features found on cellular phones of any type include display of each number before the call is sent; an accumulating call timer that records the length of all calls made on the phone; alternative long-distance dialing, which lets you use a non–AT&T service; self-diagnostic testing; automatic last-number redial; an electronic security lock; and a scratch pad memory that lets you put a phone number on the display during a call.

More Portability

One cellular phone maker has a unit with a side-mounting battery that offers one hour of talk time. A quick-slip bayonet mount means the battery can be changed in seconds. Also included in this, and in most portables, is a recharging device that plugs into a car's cigarette lighter. Automotive adapters can make a part-time home for the portable in a car. In this case, the phone might take its power from the car's battery and use an antenna mounted on the roof.

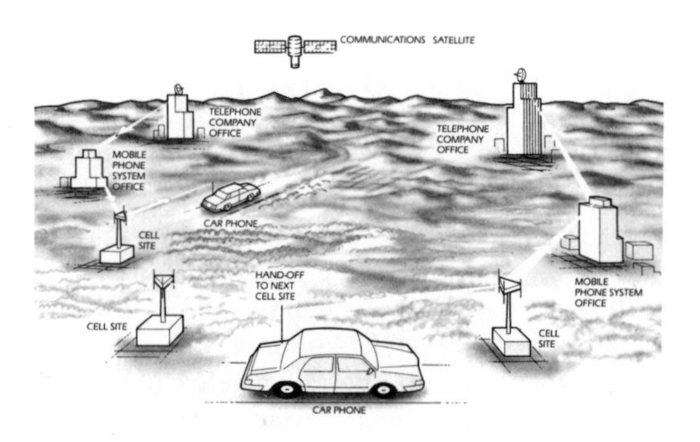

A cellular telephone system divides a metropolitan area into numerous cells, each one having a triangular-shaped antenna that relays conversations to and from the phone company's normal voice channels. As the cellular phone user travels from one cell to the next, a computer measures signal strength and passes the call to the next nearest antenna. Once mobile service transfers a call to the phone company, the conversation may be relayed by microwave, fiber optics or conventional cable. If the party at the other end is using a mobile unit, the phone company transfers the call to a cellular service. The signal then passes between the cell antenna and the mobile phone via radio waves in the 800–900 MHz band.

Alternative Service Choices

Both portables and most new cellular car phones have a feature not seen on first-generation models. This feature, called an A/B switch, reflects one of the practical limitations of cellular telephone technology. Ideally, the antenna in each honeycomb cell should be located at its center for optimal performance. But local opposition and the realities of the real estate market often make this impossible. As a result, the original honeycomb design sometimes is transformed into a mass of differently shaped amoebalike blobs.

In such a situation, a caller might be in a dead zone, that is, an area beyond the range of any antenna in the system. This problem occurs mostly in densely populated areas. Cellular phones pack only 3 watts of power (1 watt for portables), so it's possible that a cell antenna might be out of range. Fortunately, the FCC requires there be at least two competing cellular phone services in every major market. By subscribing to both services, chances are you'll be in range of an antenna belonging to at least one service. This is the reason for the A/B switch on most cellular phones.

Other Options

Because portables have only a single watt of power, not every manufacturer has rushed to offer one. The alternative is a full-powered phone built into a briefcase with its own battery pack. Talk time is generally about an hour. A cigarette lighter adapter for the car and an AC adapter for the home are included.

Another option is a class of cellular phones called transportables. Reminiscent in appearance to the crank-up field telephones carried by

This GE*MINI phone has a clip-on battery that gives about one hour of talk time. The phone is pictured in the home battery recharger.

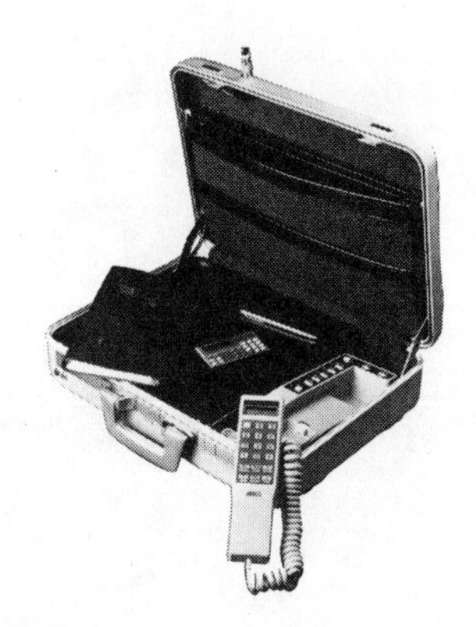

Briefcase cellular phones, such as the Audiovox model here, put out a full 3 watts and give one hour of talk time from a built-in battery pack.

the infantry during World War II, transportables are cellular phones connected to a portable battery pack. Transportables can be connected to a receptacle mounted in the car, which in turn is patched into the car's power supply and an external cellular antenna. The antennas used in car installations are superior to the stubbier versions on their more portable brethren.

Cellular technology is being extended to other areas. Specially designed marine antennas are making boat installations a reality. Other developments are opening the way toward data communications via cellular phones, providing new versatility for lap-top portable computers.

Practical Limitations

While cellular telephone technology has pushed mobile communications to new frontiers, the technology itself is still evolving. For the present, the cellular telephone is primarily a business tool sold and installed by specialty retailers and car dealers — both of whom are usually agents of either the phone company or an alternative cellular service. Its future as an off-the-shelf, do-it-yourself product is constrained by the need for expensive installation and testing equipment.

Additionally, the dealer must "burn" the phone number, local access codes, serial number and lockout codes on an integrated circuit chip (called a PROM — *p*rogrammable *r*ead-*o*nly *m*emory) before the chip is installed in the cellular phone.

On the carrier's side, problems such as dead zones remain to be solved. And as the service becomes more popular, call volume might be greater than what an individual cell can handle

Transportables can be carried in your hand or briefcase, or mounted in a car. New developments in battery technology have now extended talk time well beyond the one-hour limit.

— thereby delaying the placing or reception of calls. Today, cell size in densely populated areas is limited in practice to a 2-mile radius. Further, cell splitting requires complex signal switching (and high cost). At this point, the switching equipment often isn't up to the task. The result is interference, and sometimes inadvertent disconnection.

Digital communication techniques are being experimented with to permit a single telephone channel to carry several conversations simultaneously. Industry experts predict phones and installation will become less expensive. Factory-installed phones already are offered as optional equipment in selected luxury cars. Meanwhile, competition between the various common carriers is expected to reduce monthly charges and tolls.

The Human Factor

If cellular technology still has a way to go, so do the people who use it. While accident data are inconclusive so far, cellular phones already have sparked a debate as to whether a talking driver is a safe driver. Police in some localities will invite you to use the road's shoulder for calling if they see you chatting at the wheel. Telephone manufacturers themselves urge customers to pull over when placing or making calls.

Right now, some of the latest phones provide hands-free operation, and future use of voice activation might help alleviate safety concerns. But technology can't keep your mind on the road for you nor can it conduct your conversations. Making a sensible choice between the two is up to you.

This portable phone by Walker is ⅞ in. thick, 7¼ in. long by 2 in. wide. It's small enough to fit into your shirt pocket. It holds 99 numbers and gives 30 minutes talk time.

9
Your Own Answering Service

Many people dislike missing telephone calls. Each time they leave home — even if it's only a journey outside to their own garden — they are apt to experience a mild case of anxiety. What if an important but unexpected phone call should come while they're away? What news could they be missing?

If you're one of those people, modern technology has developed the telephone answering machine to come to your rescue. These relatively inexpensive devices can be programmed to intercept incoming calls when you're away (or just unable to answer the phone yourself). Not only will the machines inform callers that you're unable to answer at the moment, they will also ask the caller to leave a recorded message for you to hear when you return.

How They Work

Most answering machines have a simple design. After hooking one into your phone line (and plugging it into an AC outlet), you set a switch that causes the machine's electrical circuitry to react to a certain number of unanswered telephone rings. When this happens, the machine does several things in sequence.

First, it automatically releases a switchhook, which causes the incoming call to be connected to your phone. Then it turns on a prerecorded message, which is fed into the line to be heard by the caller.

The answer greeting, which you might have prepared yourself using a tape cassette inside the machine, would probably say something like the following: "Hello. This is Pat Smith. I'm not able to talk in person right now. Please wait for a beep on the line and then leave a recorded message for me. Give me your name, telephone number and the time you called. I'll try to return your call as soon as I can. Thank you."

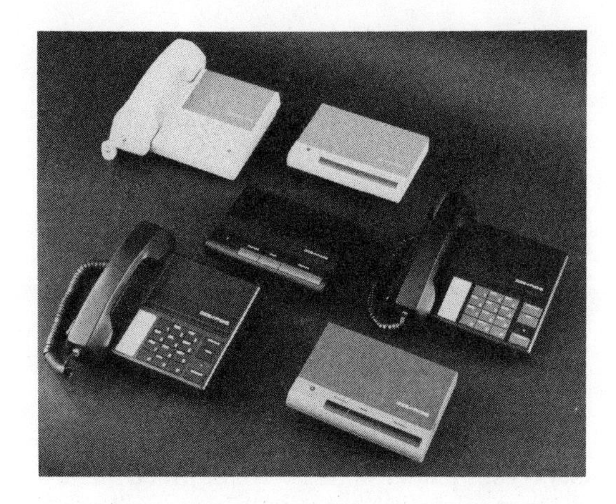

At that point, the machine will send out an electronic beep and switch from the answering cassette to a second cassette on which caller messages will be recorded.

After the caller has completed the message and hung up, the machine will stop recording, automatically rewind your answer greeting tape and then switch itself back to the "waiting" mode, all ready for the next call.

When you return home, you touch a button or two in order to instruct the machine to play back the various messages recorded by callers.

While most answering machines perform this basic sequence of tasks, some can do some additional work like replay caller messages over the phone when you call your own number from afar.

With an answering machine, people who call you while you're away can leave messages that you can listen to when you return.

Answering Machine Features

The differences between models of answering machines occur in the sophistication of operations and in the range of features.

Sound Quality. Like a telephone you buy, the quality of sound in an answering machine is an important consideration. An answering machine should reproduce the caller's voice so that you can recognize who's calling, and understand what they're saying. The machine should also transmit your greeting message clearly. When you test the machine before you make your purchase, listen to the voice quality of the messages on the tape. If you don't think the sound is natural and easy to listen to, consider another model.

Recording Taped Greetings. The simplest models likely use a single cassette tape for both playback of your own answer greeting and recording the various caller messages. This arrangement is not the best because the caller often has to face a considerable wait between the answer greeting and the recording beep. This occurs while the tape machine advances the cassette to the next free space.

An additional problem can arise if another call comes in quickly. The machine may be in the midst of rewinding its single cassette back to the start point of the answering message. Even though the machine may properly connect to the incoming call, the caller faces stony silence on the line until the rewind is completed.

Type of Cassettes. Most machines use conventional audio cassettes, but preferably cassettes without leaders so that no part of the initial message will be lost. Others use tiny microcassettes. A few even require the use of special "endless loop" cassettes, which do not require rewinding since they only go forward in a continuously repeating length of, say 15 minutes.

Most answering machines use regular audio cassettes, although some use the much smaller microcassettes and others require you to buy endless-loop tape.

Some answering machines have built-in microphones to use when you record the answer greeting.

Recording Your Greeting Message. Answering machines used on home telephones seem a bit less impersonal if you personally record the answer greeting. For that reason, most home consumers want the machine to offer an easy way to record these messages. Some machines incorporate a built-in microphone like those found on any good quality cassette tape recorder. These machines can even be used for other family recordings. Other machines are designed to let you record personal answer greetings through the attached telephone instrument.

Recording Caller Messages. Machines differ in the way they record caller messages. Good machines have what is called a VOX circuit. This keeps the recording circuit in readiness until there is a line disconnect or a long period of silence. This lets the caller leave a message of virtually any reasonable length. Less flexibly designed machines provide only a fixed recording interval of, say, 30 seconds. If the caller talks beyond that time limit, the machine cuts off and leaves the message dangling.

Number of Rings before Answer. You might be able to set the machine to answer on a specific ring when it receives a call. If you set it to answer on the fourth ring, for example, you can (or should be able to) intercept a phone message when you're home before the unit takes over by answering before the fourth ring. This can also alert friends and family members that the machine will answer if they place a call and you don't answer before the fourth ring.

Message Retrieval. Machines differ in their handling of replays of the caller messages on the cassette (or microcassette) tape. Some models require that you listen through each message from start to finish, like it or not. Most permit skipping over — and about — the tape. Some will even skip forward or backward by exactly one call.

Your ability to jump around on the tape is especially valuable when several of the callers decided not to leave messages but, instead, to hang up before the answering machine had completed its full process. This leaves a lot of useless "noisy stuff" recorded on the cassette. (Several very expensive machines automatically delete these false starts to save you from having to listen through them or to skip over them manually.)

Ease of Retrieval. It is probably obvious to you that you must keep paper and pencil handy to jot down the caller information you retrieve from the taped messages. For that reason, a machine with easy-to-operate shuttle controls of the play back mechanism can make it much more convenient to double check the names and numbers you hear. A clumsy model can cause you endless aggravation in this regard. When you're shopping for a machine, be sure to try out the playback features. (That is all the more important when it comes to choosing one of the remote message retrieval systems discussed later.)

The playback controls on the answering machine must be easy to operate so you can concentrate on writing down messages as you retrieve your calls.

Answering machines with built-in telephones, like this one, switch to a regular telephone extension when you have the answering machine part turned off.

Built-In Phones. Some of the answering machines come with a good telephone instrument built in, but most are designed to hook into your line between the wall plug and the phone you already have.

Special Features to Consider

Here are a few special features to look for when you shop for an answering machine.

Built-In Greetings. If the machine lacks a direct recording feature, you must either buy a commercially recorded message or depend on a special voice synthesizer in the machine.

Commercially recorded messages range from straightforward announcements to elaborate show-biz productions with musical jingles and celebrity impersonations.

Voice synthesizers make announcements that have been programmed by the manufacturer into a tiny computer chip. They eliminate the need for a tape recorder mechanism to make or play back the answering message. You have probably heard a voice synthesizer message from the telephone company when you misdialed a long distance call. They're certainly understandable — but, on the other hand, they nearly always sound odd and mechanical. If you're considering a unit with this feature, listen to the synthesized voice. You should, of course, be able to understand it yourself but you should also note if the voice synthesizes a man's voice or a woman's. It might make a difference to you.

Answer-Only Machines. A few answering machines can — or even must — be used strictly for answering. They do not invite the caller to leave a message. These models virtually require a way for you to record appropriate, timely messages. A typical message might say, "Hello. This is Pat Jones' residence. We're tied up right now. Please call us back after 3:30 this afternoon. Thank you for calling."

Some people feel that an answer-only procedure is really more practical than asking each caller to leave a message and a number, with a promise of a return call later. (After all, the return call may come days later — or may be to an unsolicited telephone sales pitch caller to whom you have no real interest in speaking.)

Message Alert. Manufacturers differ in the ways they have their answering machines let you know that you have messages waiting on the cassette tape inside. Some merely turn on a blinking light. Nicer models provide a digital readout that indicates exactly how many calls were handled since the last time you activated the system.

Remote Message Retrieval. As was noted previously, many models let you retrieve caller messages remotely over the telephone. This requires either a special tone generator device or the use of a tone telephone.

This tiny integrated-circuit chip is a voice synthesizer used in answering machines. It answers your calls with one or more standard greetings. They come with either a synthesized man's or woman's voice. The messages on these microchips cannot be changed.

A full-feature answering machine will give you a digital readout of the number of messages recorded, and even the time each was received.

With this device you can retrieve your messages from any phone by calling your number and sending the special tone through the mouthpiece of the phone you're calling from.

To retrieve caller messages, you dial your own telephone number. Then, after the machine emits its recording beep, you use the tone generator or the tone telephone's dial pad to send a series of coded numbers along the line. These instruct the machine to switch from its recording circuit to its playback mode, just as if you were there in person. Other coded signals can be sent remotely over the phone line to repeat or erase messages.

The remote tone-generator, which resembles a small hand calculator, has one advantage over the use of a tone telephone instrument. It can be used over any telephone. On the other hand, the device requires that you take it along whenever a retrieval is desired. That can become a bother.

Remote Call Retrieval Security. Some models offer greater security than others in the way in which call messages can be remotely retrieved. Designs that depend on single tones for remote activation are risky. A skillful mimic could actually call your line and, by "singing" the single tone code over the line, listen to a playback of your private messages. A better design makes use of multiple tones that cannot be easily duplicated by outsiders.

Call Screening. Many of the models let you use your answering machine as a call screener when you're at home. You let the machine go through its motions, listening silently on the machine's telephone extension. If the caller is somebody you really want to talk to at that moment, you can interrupt the recording and turn the machine off.

Call Forwarding. At least one manufacturer has a unit that can be programmed to forward your messages to you wherever you might be. This unit calls the number you have programmed into it and when you supply a special code, It will play the messages back to you over the other phone. The unit also lets you change the forwarding number as you move around during the course of your day's business, always forwarding your calls to the number where you can now be reached.

Time Stamping. Even though your recorded answer greeting asks the calling party to indicate the time they placed the call, many callers don't. If it's important to know when a call came in, you can buy a unit that will not only play back each message for you but will indicate on the tape (with a voice synthesizer) the time the message was received.

Call Recording. A feature you might want to look for is the one that lets you activate the machine any time you're placing a call so you can record the conversation you're having. Some of these will do it with the familiar periodic beep to tell the caller you're recording, while others do it silently. Some units will even let you use the answering machine to dictate letters and memos for you or someone else to retrieve later.

Under most phone company tariffs you are required to tell the other party that you're recording the conversation. That's what the beeps every 15 seconds do. If you opt for the silent recording answering machine, and fail to tell people you're recording their conversations, your phone service can be cut off.

Remote Room Monitor. Another remote feature found on some answering machines may appeal especially to people with unsupervised children. It allows you to dial home and listen in on the room environment from afar. If there's too much rowdy noise in evidence, you can remotely deactivate the answering machine and bring the mischief-making to a halt by reading the riot act to the child who answers your next call. This feature can also be used to monitor the integrity of your home security system.

Remote Activation. Yet another remote feature actually permits you to activate the machine by a telephone call. If you have forgotten to turn on the machine before leaving home, you can stop at, say, a handy phone booth and call your number. If you let the line ring a certain number of times, the machine is factory programmed to set itself to "on." (A casual caller could do the same thing by accident!)

Shopping for an Answering Machine

Before deciding on a model to buy, ask to see the accompanying instruction book. If it looks complicated and confusing, consider a different model. You want an appliance that makes your life easier — not a jumble of engineering perplexities.

When shopping for an answering machine with a built-in phone, be sure it is a suitable one for your home. If you choose a unit with electronic-tone dialing and live in a pulse exchange area, or haven't ordered tone service in your home, you won't be able to use the instrument to call out.

Also be sure to do a REN check before you buy. Look up the REN of the answering machine. Add it to all the other RENs on phones in your home. The REN total must be 5 or less if you want your system to operate properly.

Be sure the answering machine you buy has the notice on it that it complies with the FCC standards and lists the REN you have to report to the phone company after you install the device on your line.

10 Telecomputing Purchases

If you're considering buying a modem to communicate with other computers over telephone lines, you may have to do a little more research than just comparative shopping in a store. Lack of any real standardization among microcomputers has not made the purchase user-friendly in most cases.

In this chapter we'll tell you some of the basic things about modems you must know before you buy. We'll also let you know about some of the special modem features you might like to have. It would help to have the back-

A modem on your computer opens up a whole new world of program, data and information sources and lets you communicate "on-line" with other computerists in your community and across the country.

Adding the communications capability to your computer involves purchase decisions of a modem, the hookups and communications programs to make it work.

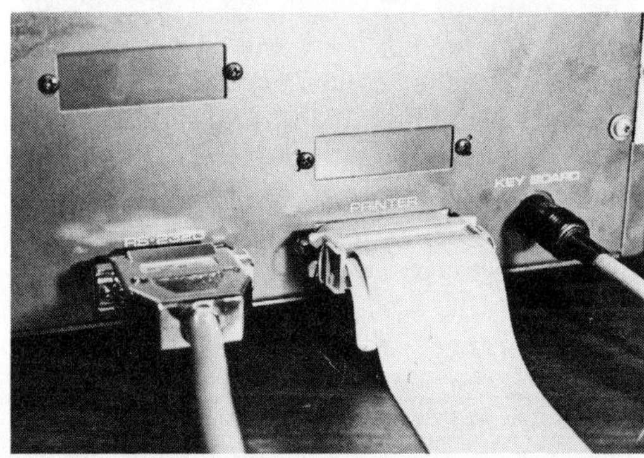

If your computer has a connector like this, labeled "RS-232C" or simply "Modem," you already have the serial board in your computer for the modem connection.

ground about how modems work that you'll find in Chapter 5.

What You're Buying

In most cases you'll have to buy a modem, an appropriate way to hook it to your computer and a communications program to make it work. Some modems come with a cable and program as part of a package. Most do not.

Is Your Computer Ready?

On some computers the serial card and RS-232C port for the modem are part of the computer's basic structure. You'll see a 25-pin connector (called a DB-25) on the back or side of your computer labeled "RS-232," "Serial Port" or maybe simply "Modem." If there is no connector — indicating you don't have the necessary serial circuits — this will add one more item to your purchase list for computer communication capabilities.

Some computers have "expansion slots" for accessories like modems. These computers will not have the connector and the modem you buy usually incorporates the serial circuits on the same printed-circuit expansion slot card with the modem circuits.

Some computers (notably smaller lap computers) have a modem already built in. On these, all you will see is the standard RJ-11 modular telephone plug to use to connect your telephone line. Since these units almost always come with communications software as well, your shopping trip is over before it began.

BASIC CONFIGURATIONS

Modems differ in three basic ways: (1) the speed at which they send out the information bits one by one; (2) whether you put them in or next to your computer and (3) how they connect to the computer. These are the first purchase decisions you will make to narrow the field down to the basic configuration you will look for in a modem.

Modem Speed

Modems send out information along the telephone lines at speeds of 300, 1200 or 2400 bits per second. This is called the *baud* rate. There are some modems that operate at speeds all the way to 9600 baud, but the most popular today for microcomputers are 2400 baud or less. You have to decide what baud rate you want for your modem.

Almost always, the faster the baud rate the more expensive the modem. Depending on your use for the modem, the increased initial cost can often be made up quickly with the savings in long-distance time charges. A 1200-baud modem can send and receive data four times as fast as a 300-baud modem — if you're sending or receiving prepared text files.

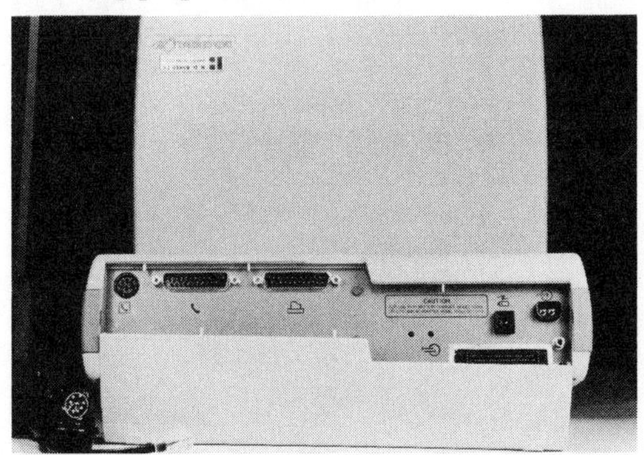

Your computer may already have a modem built in. Look for the modular RJ-11 plug on the back or side. If you find one, you can be almost positive the other end is hooked to an internal modem.

300-Baud Modems. If most of your communication will be with other computers or bulletin boards in a "chat" mode from your keyboard, there is no reason to buy any more speed than 300 baud. Three hundred words a minute is faster than almost anyone can type anyway. (That's about what communication at 300 baud works out to.)

Most of the computer data sources like The Source and CompuServe will charge lower rates per hour at 300 baud than at the higher 1200 baud. But the rates may not be proportionately lower. If you'll be using these commercial data sources, do a careful analysis of your time in use to see if the higher charge for the higher speed is going to balance against your savings in long-distance telephone line charges. There will be times, especially when sending long text files, that transmission at 300 baud is painfully slow — and expensive.

1200-Baud Modems. If a lot of your communication will be sending and receiving prepared text or program files, you should consider the faster 1200-baud modem. These modems are usually easier to use because they have more special features and better software programs than the 300-only modems.

2400-Baud Modems. Just as you can transfer a text file approximately four times faster at 1200 baud than at 300 baud, a 2400-baud modem will do the job eight times as fast. This dramatically affects the cost in terms of long-distance charges as well as the amount of time you spend on the terminal.

For most personal computer applications, the 300-baud modem provides adequate speed of communication, and these are usually priced very low.

If your computer communication requirements involve the transfer of large text files or extensive long-distance connections, the 1200-baud modem may be more economical in the long run.

The 2400-baud modems are becoming more popular with the serious telecommunicator and for business applications. This modem will transfer data eight times as fast as the 300-baud modem.

A Caveat about Speed. Remember it takes two to communicate. Modems at both ends of the communication line must be operating at the same baud rate. You can't communicate with your modem operating at 2400 baud while the other computer's modem is capable of only 300 baud.

The 300-baud modems have been around longer and are almost always the cheapest ones to buy. This, in part, accounts for the fact that they're the most popular among computerists. But the gap is narrowing among serious telecommunicators, especially as prices continue to come down for the faster models.

Almost all of the faster 1200- and 2400-baud modems also will let you communicate at the slower 300 baud (or 1200 baud). None of the 300-baud modems will let you communicate at 1200 baud. You might not find many people or services now capable of the higher 2400 baud. Check first.

Type of Connect

You can connect to your telephone lines directly, or though a device known as an acoustic coupler.

Acoustic-Coupled Modems. The early modems for microcomputers required that you dial the other computer's number on a standard telephone, listen for the distinctive sound of the other computer's signal, then place the phone in a special device that had padded, sound-insulated cups for the speaker and microphone part of the handset. This device converted the sound waves into electrical signals.

Acoustic couplers were, and still are, highly prone to data transmission errors. These are caused when extraneous noises in the room creep past the sometimes imperfect insulating cups and are mistakenly translated into false data pulses. The result is garbled transmission, often requiring that the file or message be sent again.

If most of your communication by modem, however, is going to be from hotel rooms, airports, or phone booths where modular jacks are simply not found, the acoustic-coupled modem may have to be your choice.

Direct-Connect Modems. Most modems now let you connect directly to your telephone line with the standard RJ-11 modular plug. This has solved the transient noise problems that plagued acoustic-coupled modems. Direct-connect modems also can have features such as dialing a number or answering calls without your help. If you use your computer at home, look for the direct-connect kind. They're a little

Acoustic-coupled modems pick up the signals from the telephone handset rather than directly from the lines before being converted into audible sounds.

While highly error prone, acoustic modems are the only modems you can use from places like hotel rooms and phone booths where there is no modular jack.

more expensive than acoustic-coupled modems. The difference, however, is not great enough to compensate for the frustrations of data loss and disconnects while you're trying to communicate with other computers through an acoustic-coupled modem.

Where to Put Your Modem

Modems are built either on a circuit board to insert inside your computer or they have their own case that sits on a desk or table next to your computer.

Stand-Alone Modems. Stand-alone modems are units that are designed to sit on the table or desk next to your computer. Usually these fit under the base unit of your telephone, so you don't add more clutter of equipment and cables to the area. To use a stand-alone modem, your computer must have a built-in communications port terminating in a 25-pin DB-25 connector. If there's not one there now, you'll have to add one yourself (or have your dealer do it).

Internal Modems. Many of the microcomputers today have several expansion slots in the main box that let you add circuits for many additional features — one of which is the modem card. Most of these internal expansion card modems will terminate in the standard RJ-11 modular jack on the part that sticks through the back of your computer.

Most of the newer MS-DOS computers, IBM-PCs or its clones let you add internal modems in one of the expansion slots. Older CP/M machines and some of the more compact MS-DOS computers do not. An internal modem has to be made for your computer. You can't use a modem for an IBM or one of its clones in an Apple.

If You Have A Choice. The internal modem card — if one will fit in your computer — eliminates at least one more cable plus a power cord in the cable clutter of your computer area. Internal modems are also nice if you carry your computer around between home and work; it gives you one less thing to lug around. It is usually ready for communication, powering-up with the other parts of your computer. The modem board does, however, take up an expansion slot you may want to use later for some other enhancement to your computer. The internal card is permanent, and this may be a disadvan-

Most modems sold now connect directly to your telephone line with a standard RJ-11 connector.

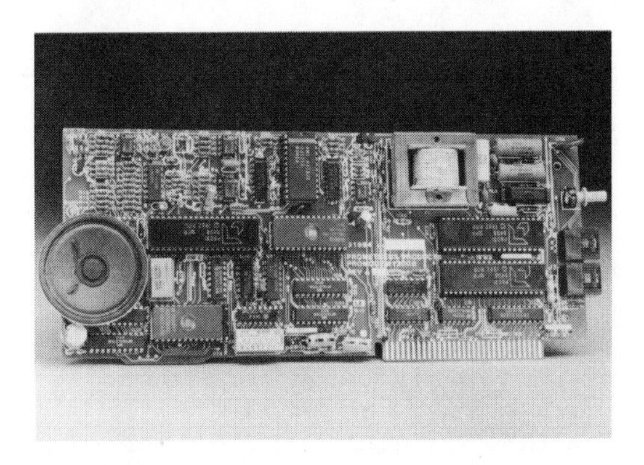

This modem plugs into one of the expansion slots in this kind of a computer, and has only the RJ-11 jacks showing on the back of the computer.

tage. Most industry experts tell us that modem standards will outlive the current generation of computers. You're likely to want to upgrade your computer before there's any real need to upgrade your modem. When you trade up, the modem card may not fit your new computer. You'll then have to replace not only the computer but the modem as well. Most stand-alone units can be plugged into any computer with an RS-232C connector.

Connecting Cables

If you have decided to install a modem card inside your computer, you don't have to worry about cables. The RS-232C serial circuits on the board connect directly to the circuits inside your computer when you plug in the expansion board. If you are considering a stand-alone modem, you'll need some way to connect the modem to your computer. In almost all cases, you'll do this with an RS-232C cable.

The RS-232C Standard. RS-232C is a standard developed in 1969 by the Electronic Industries Association. It describes the electronic characteristics and pin connections on the 25-pin connectors used. This industry-wide standard does not extend into the inner reaches of computers or of modems. There is no one standard cable you can buy to connect your computer to your modem with any confidence that it will work.

It only takes one wire and the ground to send a signal from your serial port to your modem. Another wire with the ground sends the signal from your modem to the serial port. Pin 2 on the connector, for example, is always used in the RS-232C standard to transmit data from the computer to the modem. Pin 3 is used for the return route. The other pins on the DB-25 connector are used for the "handshaking" procedures to get and keep the two talking to each other. The complete list of these connections for the RS-232C standard is shown and explained in Chapter 15.

Which Cable to Buy. Generally the computer is wired to its serial port as either *d*ata *t*erminal *e*quipment (DTE) or *d*ata *c*ommunications *e*quipment (DCE). Modems are almost always DCE. DTE and DCE devices each require cables with wires connected to different pins. If you don't know how your computer's serial port is wired, or can't find it in your computer's documentation, the safest purchase decision here is to buy the cable, at least, from the dealer who sold you the computer.

When you buy the cable, you will also have to know what kind of connections the modem expects to see on the end plugged into it. Most of the manuals with the modems will give you this information. If you're connecting a DCE modem to a DCE computer, some of the wires may have to be switched in the cable.

Check the existing RS-232C connectors on both your computer and your modem. With some combinations, you'll need a cable with a male plug on both ends because both the computer and the modem have female jacks. On some, you'll need a cable with a female connector on each end. On others, the cable must have a male plug on one end and a female on the other.

Communications Programs

Your computer has to have the proper instructions before it will work with your modem.

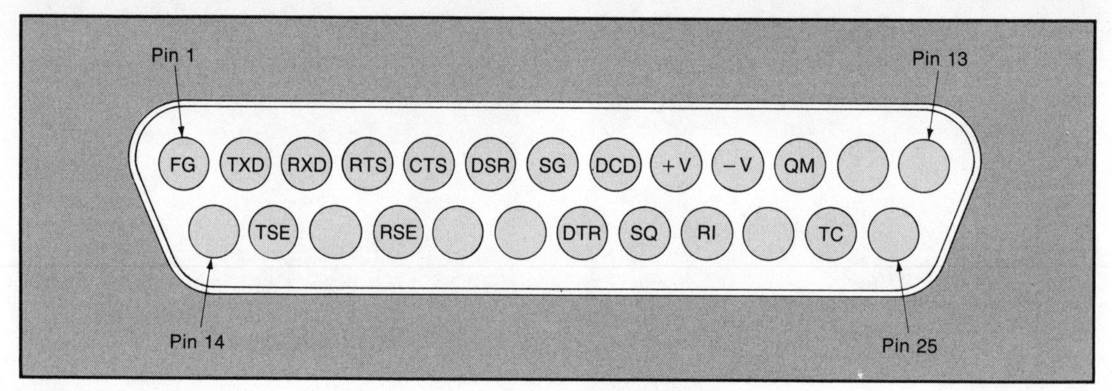

Typical RS-232C connections on a DB-25 modem plug. Not all of the 25 pins are used, and some computers may require different pin connections. The RS-232C is only a *recommended* standard, and is not followed by all computer or modem makers.

This, of course, is true with anything your computer does. Your purchase decision about a modem may well be influenced by what software it will accept and what use you intend to make of the communications capability.

You can buy the program packaged with the modem, another commercial program, or use one of the many programs developed by individuals that are available "in the public domain" from disk libraries of user groups.

Match the Program to Your Computer and Modem. Any program you buy has to work with both your modem and your computer. Most programs will have to be "installed" to match the specific characteristics of your computer, although you can buy programs already set up for computers like the IBM-PC, the Apple, the Commodore, and a host of others. If you buy a "generic" communications program, be sure it will let you use all of the features on your modem and matches the operating system of your computer. Don't succumb to a good buy on a CP/M program if your computer operates on MS-DOS.

Packaged Programs. Many times the modem will come packaged with a communications program specifically designed to use all of its features. Almost always you'll pay extra for this program, even though it looks like part of a "package" buy. The program provides the correct settings for your modem, but you'll have to set it for the protocol your specific computer understands. The protocol jargon of such specifications as "full duplex, with eight data bits, one stop bit and no parity checking," will be clarified for you in Chapter 15.

Generic Commercial Programs. Most vendors, in efforts to keep prices looking low, will not include the communications program with your purchase. Be sure you can buy the communications software package that can make the modem work with your computer and let you use all the features of your modem. For example, if your modem has an autodial feature, you're not going to be able to use it unless the software program supports it. At the same time, be sure the program will match the serial port characteristics of your computer. The best advice if you're not sure is to ask a knowledgeable clerk, getting a specific statement that the program will work with with Brand X modem in Brand Y computer.

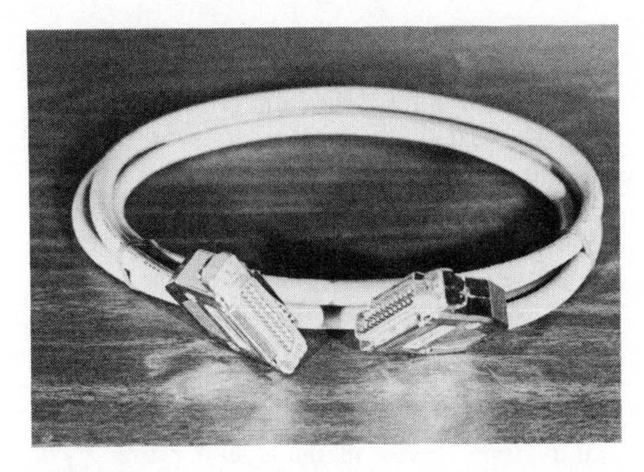

RS-232C cables come wired to connect a data terminal (DTE) computer to a data communications (DCE) modem. You can buy cables with pin arrangements to connect DCE to DCE or DTE to DTE if your computer and modem require these connections.

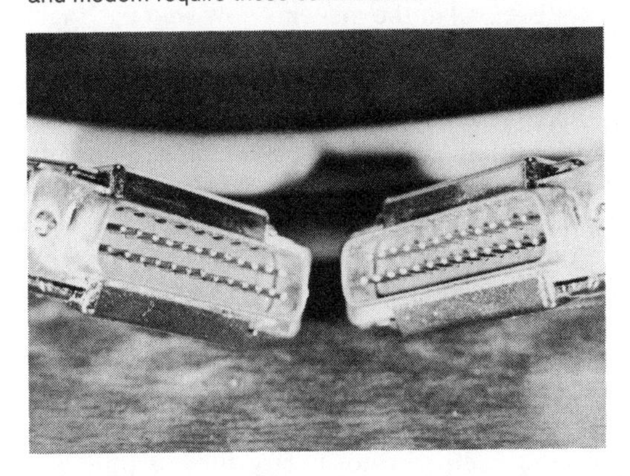

Some RS-232C cables have male connectors on both ends, while some have a female connector on one end. Check what kind of connectors are on the back of the modem and computer before you buy a cable.

Public Domain Programs. There are also communication programs written, developed and distributed by user groups. The oldest and still most widely used is the current version of Modem7 written for the National CP/M User Group. Programs like Modem7 are in the public domain and you are usually charged only the cost of a disk and maybe a modest contribution to support the user club library operations. These are good communications programs. Many of their commercial counterparts have their roots deep in these public domain efforts.

With most of these communications programs, you will have to do some modifications to match the characteristics of your computer and program. This will require that you know more than just the basics of programming and making changes in .COM or .EXE files. If you get lucky, you can get the program from a user group that specializes in your brand of computer, where some more dedicated member has installed the public domain program for the computer and all you have to do is take it home and run it.

Hayes Compatibility. Many modems advertise as "Hayes compatible." This means they respond to the same set of codes as the Hayes Smartmodems™. Since the Hayes Smartmodems were an early sales leader in the modem field, many software packages generate commands that Hayes modems understand. The claim of Hayes compatibility should mean a modem will be supported by a large number of communications software packages.

There are degrees of Hayes compatibility. Not even all of the modems now sold by Hayes are completely Hayes compatible.

Shopping for Software. Look for a dealer that sells both the modem and communications software. Most of the time you can be assured that you'll get a matched pair. If you don't, you have a good case for return or exchange. It's a good idea to have the modem demonstrated with the communications software so you know that the combination will do what you want it to. For example, some will only allow you to transfer ASCII files (like from your word processor or spreadsheet) and not binary files of programs you write and want to exchange with others.

SPECIAL FEATURES

All modem functions are controlled by the unit's "firmware." Firmware is software that is permanently written to the integrated circuit memory of the unit and cannot be changed. A "smart" modem will do many of the functions for you, like dialing out or answering a call. "Dumb" modems don't. While some dumb modems can be made a little smarter by a good communications software program, most lack some of the special features that make communications between computers easier and better. Listed below are some of the special modem features of smart modems you may want to consider in your purchase decision.

Autodial

Some modems have a memory that lets you store the phone numbers of some of the other computers or data services you call frequently. With one or two simple commands, the modem finds the number in its "directory" and then sends the correct dual-tone dial signals over the phone lines to place the call. You never have to touch your telephone dial yourself. Most modems with autodial will also have a redial feature. If the computer's line was busy on the first try, it will keep redialing the number at regular intervals — say every 3 minutes — until the line is free, or until you stop it.

You can use this autodial feature on some modems to place regular voice phone calls by picking up your regular telephone without going through the modem and computer with the voice signal.

Power Supplies

All internal modems draw their power from the computer's power supply. The purchasing advice here is a caution about overtaxing the power supply capacity with too many expansion boards. At some time, as you add features to this kind of computer, you're going to have to think about a bigger power supply, too.

External modems are usually powered by the familiar "battery substitute" box that plugs into the AC wall outlet and sends low voltage to the modem on a thin, two-conductor wire.

There are an increasing number of modems, however, that come with a built-in power supply. These have just the larger cord that plugs directly into the AC wall outlet.

External power supplies can be replaced; internal power supplies must be repaired. External power supplies limit where you can put the modem, usually no farther away than the length of the cord from the wall unit. With an internal power supply, you can extend your reach with a normal AC extension cord. Often, internal power supplies are easier to connect to computer surge protectors than are the battery eliminator boxes of the external power supplies.

Some modems have an external power supply, connected to the modem in much the same way you hook up a portable radio or tape recorder to AC power to save the batteries.

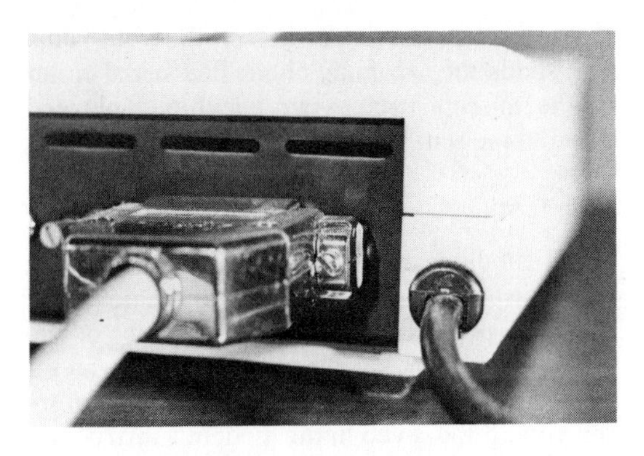

This modem has its own internal power supply. It connects directly to a wall outlet or to a surge protector with your other computer equipment.

Auto Answer

Modems can also be left on, connected to the phone line and computer, set to answer the phone when it rings and then exchange data with another computer completely unattended. This feature lets you receive (and sometimes send) files while you're away doing something else. It can also be very annoying to callers to be greeted with the high-pitched squeal of computer communication when they dialed your number for a voice conversation. Most people that use this feature have a separate line into their homes dedicated to computer use only. Some telephone directories now let subscribers include separate numbers for "voice" and "data" in their alphabetic listings.

DIP Switches

Most of what a modem does is controlled by either the software program or tiny switches on the modem itself. These are called DIP switches because they fit into the same DIP-type socket on a circuit board that integrated circuits use. They are small and rather delicate. Usually about 10 on–off switches are crowded together on a chip-type unit only about ¾ in. wide.

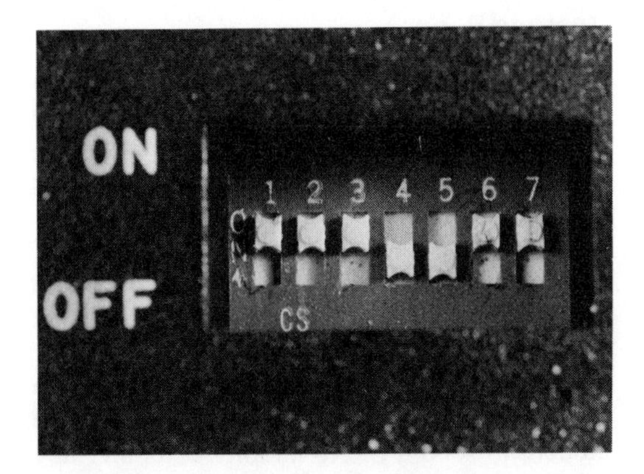

These tiny switches fit in the same space as an integrated-circuit chip. They are used on many modems to set certain functions as required by your computer or software program. These do not normally have to be changed during operational use of the modem.

Some modems have the DIP switches behind the front panel, hidden but in relatively easy access for you to use. Others put them on the bottom of the modem, making you remove the phone and twist the wires to change the settings. A few locate them on the back. Some don't have DIP switches at all, but depend completely on a communications software program to change the settings.

If you're going to use the software communications program to change the protocol parameters and other functions of these DIP switches, it really doesn't matter where the modem maker has put them. You probably won't need to touch them after the initial setup. If you're going to have to reset the switches manually and frequently, you most certainly want them conveniently placed.

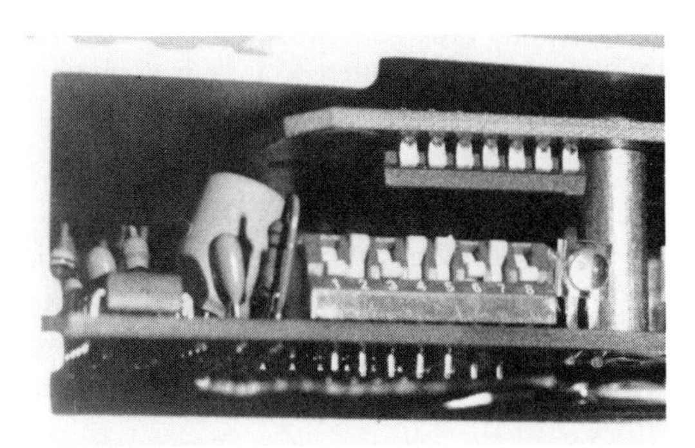

The DIP switch is often located behind the snap-off front panel of the modem.

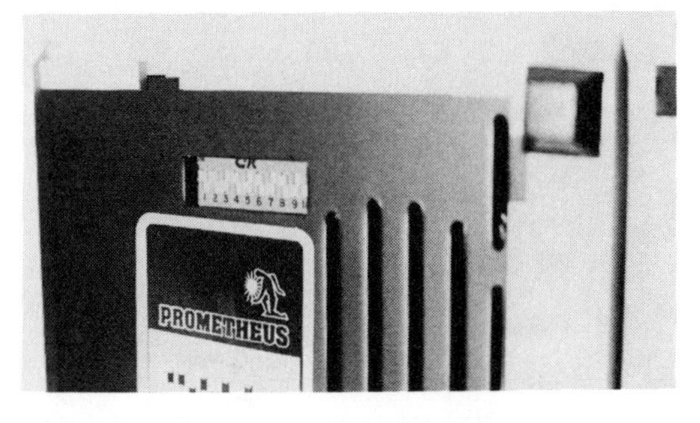

Other modems will have the DIP switch on the bottom, making access to it a little less convenient.

Status Indicators

Most modems have a series of panel lights that tell you what the modem's doing. The light labeled TD, for example, lights when your modem is *t*ransmitting *d*ata. The RD lights when you're *r*eceiving *d*ata. CD tells you when a signal (*c*arrier) has been *d*etected from another computer. This mean's you're hooked up and ready to communicate. The OH light means your modem is "*o*ff the *h*ook" and online ready to communicate.

At least one modem manufacturer offers a digital display accessory that will indicate in plain English words what's happening at every step of the way.

If the modem you're considering doesn't have at least the basic minimum TD, RD, CD and OH indicator lights, you may have trouble monitoring any problems with your data communications. Once your computer screen shows you're communicating, you may never even notice the indicator lights on the modem.

Voice–Data

There will be times when you will be in voice contact with another computer operator, then want to switch to data transmission. Virtually all software programs will let you do this through commands entered on your keyboard.

Most modems have these panel lights to tell you what the modem's doing. They give you a quick visual check on proper operation during a session.

This modem tells you what's happening with a digital display. Since this is often an accessory option, these modems also have the more conventional status lights.

A few modems have a switch on the front panel that sends the incoming phone line signal either to the modem or to your telephone, plugged into the modem.

Self-Test

Some modems go through an internal testing procedure every time you turn on the power. If there is a problem in one of the functions or circuits, the modem signals you to fix or correct it before you continue. Very few manufacturers tell you how — even in the modem's instruction book. If the self-test diagnostics procedure does reveal a problem that you can't correct, you'll have to take the unit in for repair. The self-test feature does, however, give you the confidence to begin your communications when the system checks out okay. It's a feature that avoids a lot of frustrations.

Real-Time Clock/Calendar

Many manufacturers now offer, as part of the modem or as an accessory to it, a real-time clock and calendar. Some of these clocks are backed up by a battery to continue keeping time when the power to the modem is off. Others require you to set the time and date every time you power up the modem.

The clock/calendar is handy if your software program is configured to use it. Incoming files can be electronically stamped with the date and time you received them. Most of these clocks will give you a running readout of the time you're connected to another computer — a handy feature if you're calling long distance.

Message Buffer

An accessory item to several modems now is a message buffer. These buffers let your modem receive and send files, even when your computer's turned off. Depending on how much you upgrade the storage capability of your computer, you can store incoming or outgoing messages in one of these buffers up to a maximum of 512k (the minimum buffer size is usually 64k).

When this message buffer is combined with the real-time clock/calendar, you can set the modem to send a file to another computer any time of the day — or night when the long-distance rates are lower. At the same time, this lets you receive messages from other computers much like your own personal electronic mailbox.

A message/communications buffer like this is an accessory item for several of the more sophisticated modems. It lets you set up an "electronic mail" system where you can store incoming messages or hold outgoing messages until it is more convenient (or less expensive) to send them. With stand-alone modems, these buffers continue to work as long as the modem's turned on, receiving and sending messages, even with your computer turned off.

Printer Interface

Some modems also use this message buffer as a printer buffer, allowing you to send material directly from the modem to a serial printer without going through your computer. It also gives your computer added capacity for printing long documents by storing the ASCII text in the buffer for printout, freeing your computer for other tasks while the printing is being done.

OTHER THINGS TO LOOK FOR

Besides the optional special features you may or may not want on your modem, there are some features it must have and others that will make your life a lot easier.

FCC Certification and REN

Your modem must have the FCC certification and should specify the ringer equivalence number (REN) both on the unit and in the instruction booklet. You have to report the REN of a modem to your telephone company just as you do when you add another telephone in your home. The REN will also tell you if you've got your home telephone system overloaded. If your REN total is more than 5 after you add your modem to the phones in your house, you will have trouble getting and placing calls. It can also mean the phone company can cut off your service if they determine you're interfering with the service of others.

Instruction Booklet

Most modems come with pretty good documentation. Most of this documentation, however, is written for both the devoted computerist who knows what's happening and for the novice who doesn't know a baud from a bit. Efforts to reach both audiences are not always successful. Be sure the instruction book of the unit you intend to buy is one *you* can understand. They should all have a glossary to define the terms used in a way people can understand, but most don't.

A quick reference card that can be removed from the manual and kept by the computer to give you an overview of what to do is a handy item to look for in the documentation. It's better than trying to thumb through 144 pages of information when you're trying to remember a specific command during a data transfer session with another computer on a long-distance call.

Telephone Connections

Modems will have either one or two RJ-11 mod-

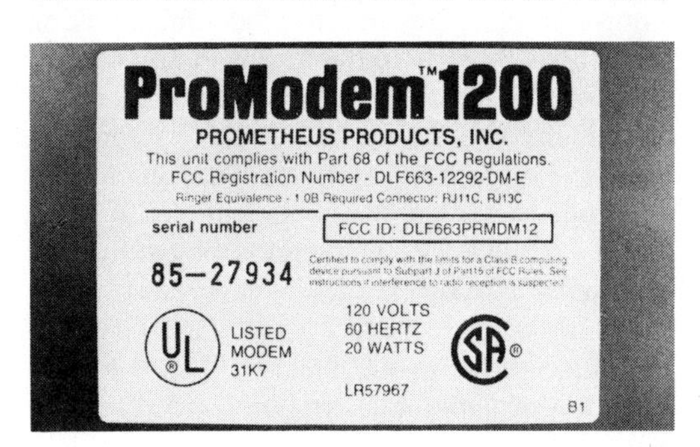

Any device connected to your phone lines, even a modem, must be certified to comply with FCC standards and have a REN, which you must report to the phone company when you hook it up to your line.

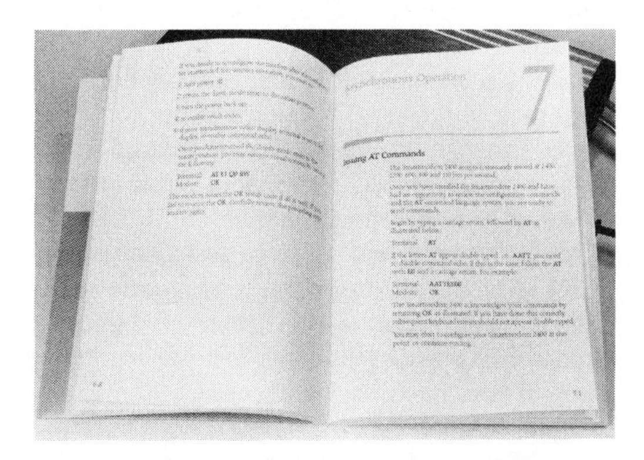

Documentation for your modem is an important factor to look for when you make a purchase decision. No modem, no matter how sophisticated, will be any good to you if the instructions don't tell you how to make it work.

In this modem, a telephone can be plugged into the modem along with the cord to the wall jack. When the modem's not in use, the telephone can be used just like any other extension phone. Most also let you switch between data communications and voice when the modem's on, cutting out the phone when data transmission is taking place.

ular jacks, usually on the back. Units with two modular jacks let you connect the modem to the telephone line and your phone to the modem. When the modem is turned off, the jacks "loop through" inside the modem, hooking your phone directly to the line as in any normal telephone connection. When the modem is turned on, you can usually select through a switch or through the software whether the signal from the telephone line will go to your phone for voice communication or to the modem for data transfer.

On those modems with only one RJ-11 jack to connect to the phone line, you'll have to provide your own modular Y-type connector if you want to use a phone on the same outlet with your modem.

Surge Protectors

It makes good sense to feed the modem its AC power through the same surge protector you use for your computer. Most let you hook up more than one device. Sudden voltage spikes or power surges in the electrical lines can destroy the integrated circuit components in a modem as fast as they do in a computer.

It also makes good sense to add a surge protector to your telephone line. Telephone wires outside your home are as prone to lightning strikes and power surges as are the electrical wires. These inexpensive accessory items usually plug into an AC wall outlet, with only the grounding lug of the three-prong plug made of metal. You must have a properly grounded outlet if this type of surge protector is going to give you any protection.

Surge protectors should be used on both the AC power line and on the telephone line connected to the modem. Destructive voltage spikes and power surges can invade and destroy your computer through telephone lines as well as through the power lines.

WHERE TO SHOP

The first place to look for a modem you know will work with your computer is at the store that sold you the computer. In most cases, you'll get better dealer support and have some return to exchange privileges if it doesn't work.

Mail-order firms are likely to be substantially cheaper, even when you add the cost of postage and handling. You're not likely, however, to get much help if you don't understand something or if you've hooked something up wrong. Most of these mail-order houses have an 800 toll-free number to place your order. When you call and before you place the order, ask about the return policy if the unit is defective or damaged in shipping. Don't ask if they'll let you return the unit if you just don't like it. Most dealers, retail or mail order, don't like you to do comparative shopping at home with their merchandise inventory. Some will let you return a defective or damaged unit for an exchange — maybe even a refund. Others will insist you return the unit to the manufacturer or to an "authorized service center." Avoid these if you can.

Many computer user groups will get together on a cooperative group-buy, getting a special lower price from a manufacturer or dealer because of the volume sale. You're likely not only to get a good bargain with this arrangement, but also a resource bank of other people to contact if you don't understand something or can't get something to work.

11 Phones Where You Want Them

In today's deregulated communications environment, you can design, purchase and install your own telephone system inside your home. Anything on your side of where the telephone company terminates its lines is, in fact, your responsibility. In most localities, you can have the wiring inside your home done by the telephone company — but at a cost. There is both a challenge and a satisfaction in doing it yourself. If you follow these directions, you'll end up with a communications system in your home tailored to the needs and desires of your family.

Home telephone systems can be installed using only modular cords, extensions, junctions and adapters. This may be a necessity in apartments where a more permanent installation is not allowed or desired.

PLAN YOUR WIRING

All of the plans for telephone wiring in your home will begin at the phone company's terminal. Before you start stringing and cutting wire, however, it's sensible to take the time to plan exactly where you want the phones now — and where you might want them in the future. In only a few short years, the nursery you're adding on to your house now will be occupied by a teenager with a voracious appetite for phone calls.

Types of Wiring Installations

Consider first the kind of wiring installation you want to do. You can plan to run modular extensions from existing outlets; run extension or bulk wiring along the baseboards or moldings; or you can do a completely hidden in-the-wall wiring installation.

Modular Extension Cords. Both to plan and to execute, it is easiest to use prewired modular extension cords to extend the reaches of your telephones from existing modular jacks. If yours is an older home, you may have to modify older jacks to accommodate the new modular cords, but that is not a difficult job.

Baseboard Wiring. In many older homes, original telephone installations by the phone companies were done by running small, round multiconductor cables stapled to baseboards and moldings throughout the house. This is still a good installation technique if you don't want the temporary look of modular extensions running all over and if you can't tackle a more permanent job of hiding the wires.

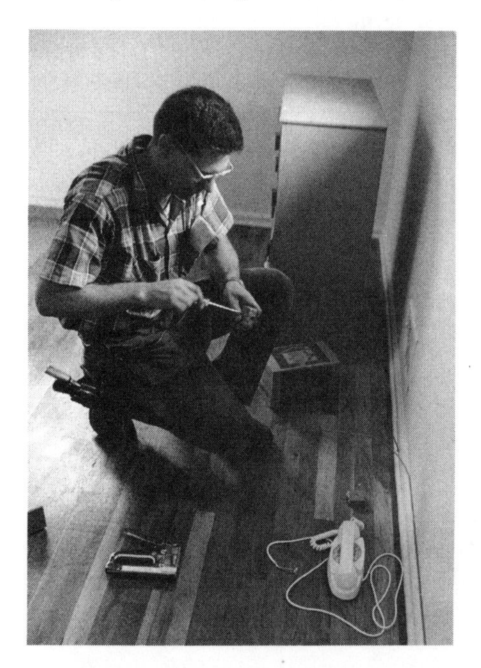

Bulk color-coded wire can be stapled to moldings and trim using surface-mount jacks and junctions. This type of installation requires less carpentry work than does hidden writing. Almost all phone company installations used this method in the past.

In-the-Wall Wiring. For more permanent and less obtrusive wiring, you can plan to run the wires inside the walls. This is, of course, easier in new construction or where your remodeling project exposes the studs. It is not impossible, however, even in older homes with established walls you don't want to rip out.

Locate the Junction Box

Your home telephone installation starts at the utility's junction box — with its protector tap-on devices inside. You'll find the junction box where the incoming telephone line terminates at your house. Sometimes it's outside the building; more often it's inside. Look for it in the basement, the attic or in a utility area.

Phone companies install a network interface in many homes. This is a tiny integrated circuit that protects the phone network from any harm by equipment you install. It terminates in a modular jack rather than screw terminals to which you connect your home wiring.

There may also be just a terminal junction where your responsibilities for home wiring take over. These have two terminals to attach the telephone wires and a third terminal that the phone company attaches to a water pipe or other suitable ground in your home.

If you live in a multifamily dwelling and can't locate your terminal for certain, you may have to ask the utility company to send out a technician. You will be billed, but it's better to be

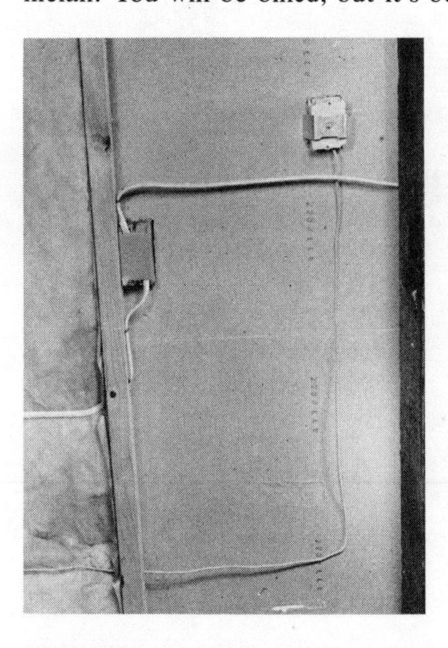

Many do-it-yourselfers will opt for a home telephone system with wires completely out of sight inside the walls and using flush-mount jacks and junctions. This type of installation requires careful planning and carpentry skills.

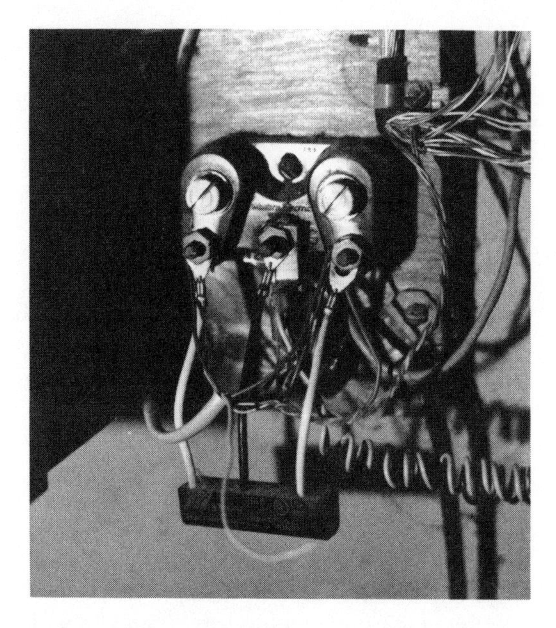

Your home telephone system starts where the phone company's wires terminate at or in your house. The phone company's wires may come into the utility area near your electrical fuse box. The lines may also end on the outside wall of your house.

safe than sorry. Connecting your lines to a neighbor's line by accident is likely to prove more than embarrassing. Of course, if the neighbors are willing, you could do a quick disconnect of the wires to see if the phones in your house — or the neighbors' — have been cut off. Be sure to replace the wires properly if the tap-out isn't yours. Identifying the wires at the junction boxes for future reference is also a good idea.

No matter what kind of termination the phone company has made on the lines coming into your house, you should leave their connections and equipment alone. Many terminal blocks and junction boxes will have large resistors between two of the terminals. The wires for these resistors should never be moved. Some may have fuses or other protective devices, which are also to be left alone. Under no circumstances should you change the way the utility's yellow ground has been attached. That could jeopardize your whole system — not to mention your home.

Telephone Wiring Cautions

There are some precautions you should take before you progress too far in your planning. According to the phone company, a person installing a phone should be sure the installation

Be sure you know which terminal is your line if you live in a large condo or apartment complex where the phone company terminates all lines at one place.

complies with local building regulations and the National Electrical Code. Don't plan to place telephone wires in pipes, conduits or compartments containing other electrical wiring. Never place telephone wiring near bare power wires or lightning rods, antennas, transformers, steam and hot water pipes, or heating ducts.

Before you fasten any wire to metal surfaces — siding, recreational vehicles or mobile homes — be sure no hazardous voltages are present on the siding or other conducting surfaces. And, you must never run wiring between structures where it may be exposed to lightning. Further, avoid damp locations or any place where wiring lets a person use a telephone while in a bathtub, shower or swimming pool.

Never use telephone wire (yours or the phone company's) to support objects. Be sure the wire is protected by electrical tape when it runs across gratings or other rough objects.

Where Do You Need Phones?

Planning and installing a home telephone system is a job you want to do only once.

Do a Family Survey. Do a very careful survey of your house to see where you want phones. Consult family members — especially the children — about their preferences. Observe what rooms the family spends most of its time in when phone calls come in. Carry your observations one step further and study where people usually are in these rooms. It will make a big difference when you finally decide on what wall to install the phone outlet. Watch your family's habits. Teenagers may prefer to talk on a phone while dangling over their beds rather than sitting at a desk. There will also be times when they — and you — want a place for a private conversation.

At one time, the telephone was located in the living room. Way back then, it was a family event to make or get a phone call. Now, the most convenient rooms in the house for phones are the kitchen and bedroom.

Kitchen Phones. The center of activity for many families is the kitchen. A kitchen phone becomes the focal point of a home communications or finance center. It becomes more than a convenience when you're in the middle of an extraordinary culinary feat and need instant communication with someone across town or across the street.

Bedroom Phones. Telephones are now as common on the bedside nightstand as are alarm clocks and lamps. No longer do you have to stumble through a darkened house to answer a ringing phone; for some reason, emergency phone calls seem to happen more frequently in the middle of the night! Some people have even gone so far as installing separate "his" and "hers" lines from the phone company with an individual phone on each side of the bed.

Phones in the Study and Family Room. The study, den and family room are obvious choices for a telephone. Both family activities and individual privacy can be accommodated by a convenient phone outlet.

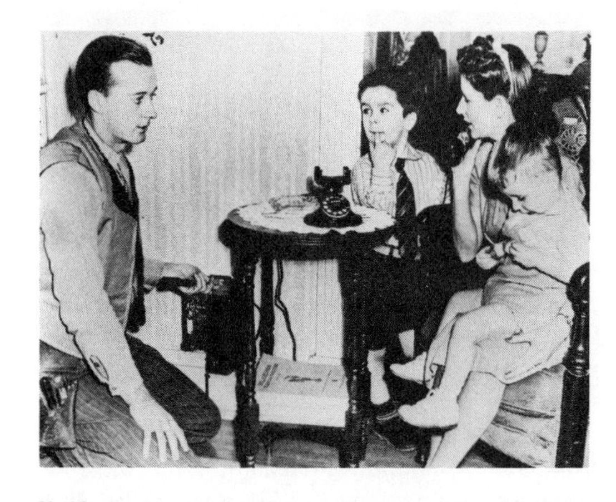

Not long ago, it was a family event to make or get a phone call — especially a long-distance call!

A kitchen phone is both a convenience and a necessity for family communication and the culinary artist.

Telephones are now as common on the bedroom nightstand as are alarm clocks and reading lamps.

Workshop and Basement Phones. The place you may think least about installing a telephone — until you're there and one rings — is your workshop or basement. A workshop phone gives you the convenience of not having to shut everything down for a call. It will also save any hassle caused by dragging the sawdust into the living areas of the house.

Computer Communications. If you have a computer in your house, plan now to install a phone line near the computer work area —

A phone in the family room can accommodate both family activities and individual privacy.

whether or not you have a modem to communicate with other computers now. Besides its use with a modem, this phone location can save you the frustration of trying to save the file you're working on quickly enough to run across the house to answer an impatiently ringing phone.

Phones by the Pool, Patio and Garden. Running phones outside can be tricky. Best to leave the installations at the pool or patio for a professional. You can, however, plan to install a bell outside to let you know when the phone is ringing in the house. Most of these hook up with a modular RJ-11 plug or are hard-wired with the same color-coding scheme as other phone equipment.

A workshop phone keeps you in touch without interruptions in the middle of a craft project.

A phone by your home computer is a wise decision even if you don't have a modem now for communication.

Locations to Avoid. Be careful about installing phones in your bathrooms, as most of the luxury hotels seem to do today. While the voltages and current in a telephone system are low, they can still give you quite a jolt when you are using a shower, sink or any other plumbing fixtures that give wayward voltages a path to ground through you.

Outside installations can be tricky but bell extensions can summon you to your inside phone.

Locating a Phone in a Room

The next step is to plan carefully where the wires will be run for the phone outlets in each room. There are several important things to keep in mind that will make your job easier and the system more efficient and convenient.

If you are running new wiring, either in the walls or along baseboards and moldings, remember that phone connections along inside walls of a room make your wiring easier. This will also make runs between connections shorter. For example, if your plans indicate a phone outlet in the middle of a wall dividing two rooms of your house, each of which will have a phone, locate the phone in the next room on the opposite side of the same wall. This will cut down the wiring distance between the two outlets to the 5-in. thickness of the one wall. If, on the other hand, you decide that the phones should be on the far walls of two adjacent rooms, you'll have to plan on a much longer run for the wires between the jacks.

An important consideration in planning where you'll put the phone jack in a room is furniture placement. Normally, you'll want to locate the jack no farther than about 4 feet from where a desk phone will be placed. That will give you just about enough slack with a 6-foot cord that goes to the jack on a phone you buy. While you can buy longer extensions, careful planning of the location of the jack can save you this extra cost. It can also save the sometimes unsightly tangle of wires you'd find if you had to substitute a 15-foot extension to reach a jack

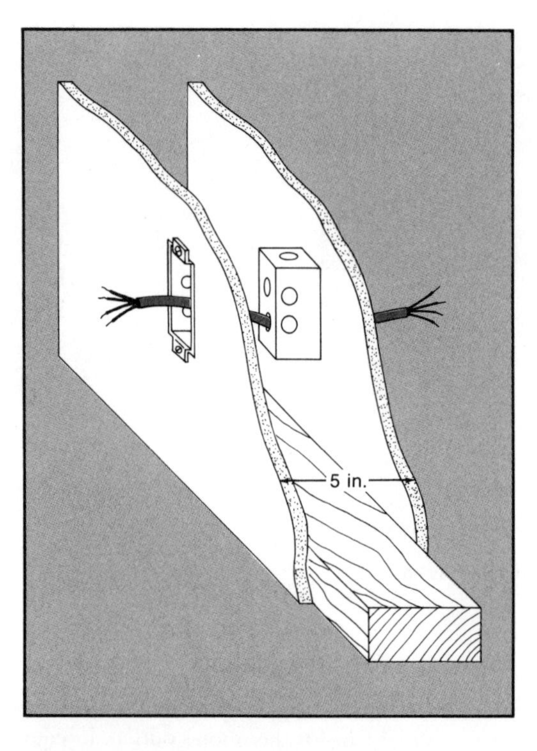

The length of phone wire between outlets in two adjacent rooms can be very short with careful planning.

you put 7 feet away from where you wanted the phone.

Your study of the family's activity habits within the room you're planning is critical here. If family room activities center around a built-in-bar, it may not make much sense to locate the telephone next to the dart board.

You should decide now whether the installation in the room is going to accommodate a wall phone or a desk phone.

You have the advantage with a wall phone installation of locating it just where you want it without considering where a desk or table on which you're going to place the phone is located in the room. On the other hand, you can't hook an extension to a wall phone and move it to a more convenient location later. Long coiled cords between a wall phone instrument and its handset are predestined to become a tangled mess within a very short time.

Draw Your Wiring Plan

The next step in planning where you'll install the wires is to draw a rough scale floor plan of the rooms in your home. Then draw the location of the telephone wiring on the plan.

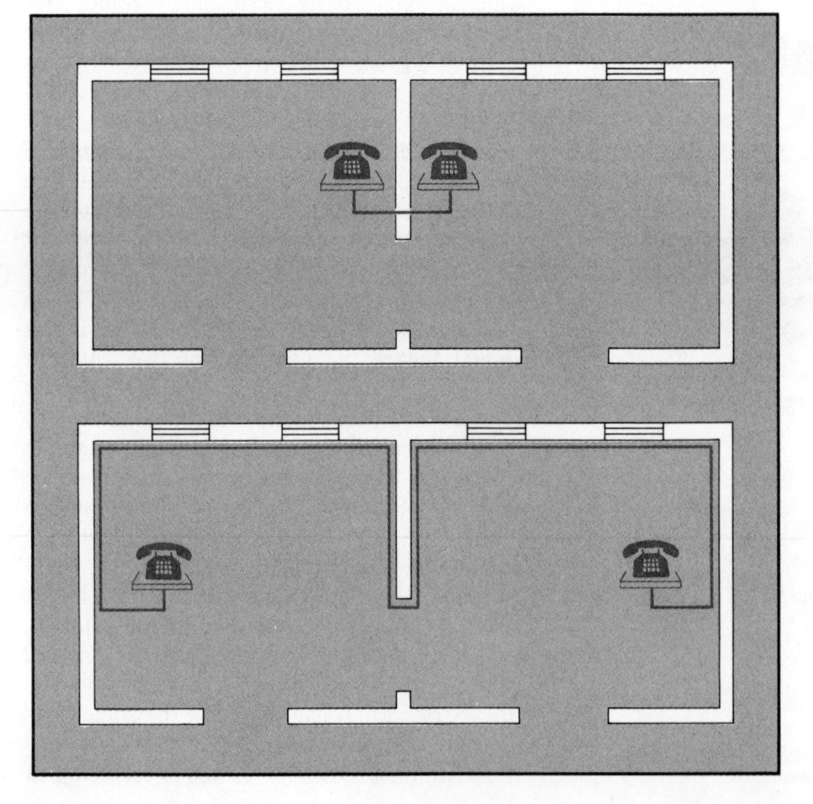

Phones on opposite walls in two adjacent rooms require longer wire and more carpentry work.

If you're planning a phone system for a new home or for a major remodeling job in your existing home, you'll have an accurate set of plans from the builder. You can also use the floor plan of the room layouts that normally is included with real estate listings.

If you're installing your phones in an older home, you'll probably have to draw the plan yourself. Exact scale is not essential. Don't, however, let measurement errors compound, leaving you several feet short when you string the wires. Some of the critical measurements (these must be accurate) are the distance from a corner to a door or window jamb, the location of electrical outlets or plumbing fixtures, and the location on the wall where you're going to install the phone jack.

It may also be a great help later on if you identify, in each room, what kind of access you have below the floor and above the ceiling of that room. Many times, in-the-wall wiring through an unfinished basement and up the inside walls can be run more easily than trying to make your runs through the walls from the room itself.

Now locate the furniture (in rough scale) on the floor plan. This is a good step to follow even when you're installing extensions rather than permanent wiring for your system. You don't want cords to get in the way of other activities. You want to plan now to keep cords where they can't get caught up in feet, chairs or the vacuum cleaner. At the same time, don't make your plans so tight that you have to run cords under carpets or out in the open areas. That's a sure way to create problems later on.

Locate Your Wires on the Plan

After you've located the position of each phone outlet on your scale drawing, plan how you'll route the connecting wires between them. This is going to be a preliminary plan at first, since you'll have to check out the location of power

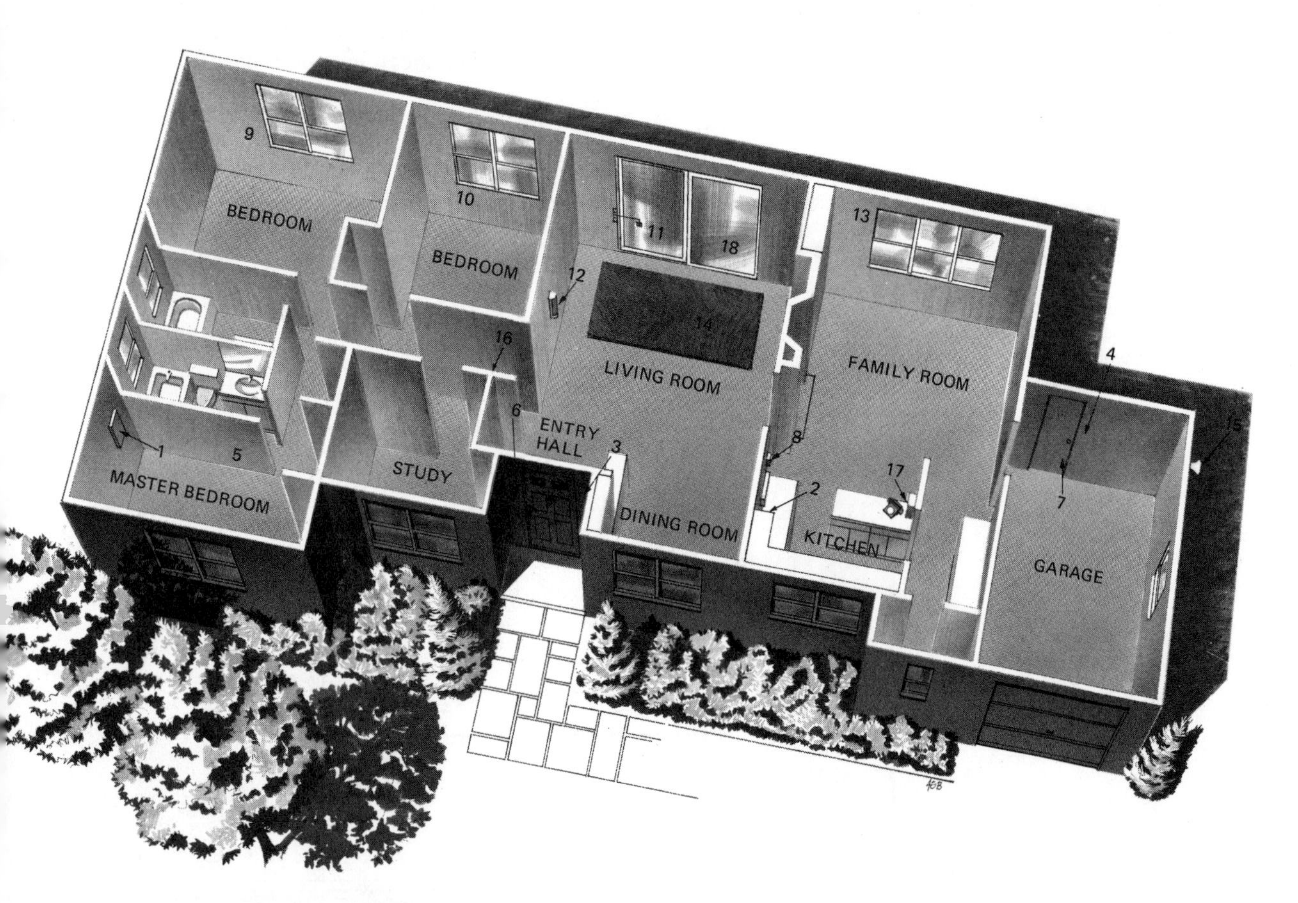

Planning a home phone system begins with a rough floor plan to locate where you want phones and to plan wiring.

outlets and any plumbing fixtures along the run. You'll also have to get a rough estimate of the total length of the wiring for phones in your house. Measure this from the telephone company's terminal box to the last phone in your chain.

According to AT&T, the maximum distance you can run a wire from the network interface or junction box is about 250 feet. This is the limit before line resistance and voltage loss become too great to operate the phone. You should also not plan to install more active phones than the line can handle. The phone company uses a system that supplies 48 volts and between 20 and 35 milliamps of current. This provides excellent service on one phone. As you add extra phones to the line, you are dividing this current and voltage further, and the quality of the service deteriorates. The REN on each phone gives you a fairly accurate gauge about the number of phones you can install. The total RENs on all the phones in your home should not add up to more than 5. This usually means about six or seven phones, but it will pay to do the simple arithmetic.

Chain Wiring Installation. Running your telephone wires in a chain means that you start at the telephone company's terminal and with a single cable you hook up the first outlet, then run the wires from that outlet to the next along your plans. Each location along the system depends on the connections at the last jack or terminal block.

This is usually the easiest and most economical way to run the wires for your telephone system. There are times, however, when you just can't reach the next outlet from the last one you installed. In these cases, you should consider cluster wiring.

Cluster Wiring Installation. The cluster wiring means you'll run a separate cable from the telephone company's junction to each phone outlet you want to install. Often times you can get under the maximum distance of 250 feet by running separate lines for each outlet if the phone company's terminal is located in a straight line up to the rooms where you'll want phones. This installation may require more carpentry work than chain-type wiring. It will also require a larger junction to feed each room, or you will have to link several junctions together to get the number of wire connections you'll need.

Using Wire Junctions. You can combine both

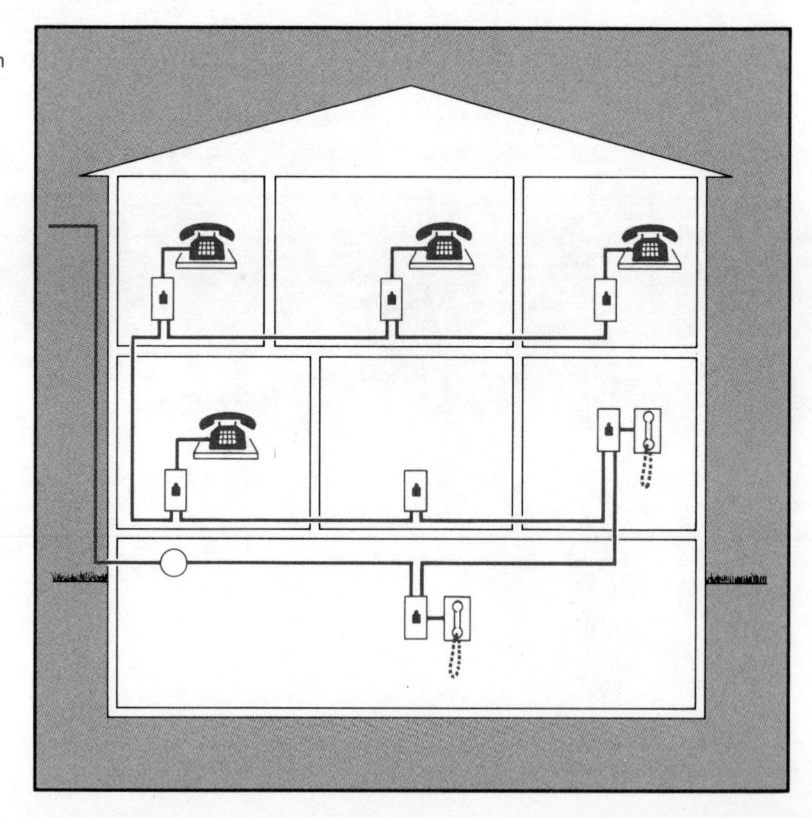

This is an example of chain wiring in a home. Each outlet in turn is fed with wires from the previous outlet. This type of wiring requires shorter wires between each phone location, but the total length should be kept under 250 feet.

chain and cluster wiring in your installation. You'll find that you can run your line using the more economical chain-type routing until you get to a point where it makes more sense to split the line to go to two rooms rather than run a separate longer wire from the phone company's junction. Where this is done, you can buy wire junctions that will let you connect anywhere from two to four branches off a single feed wire.

It is better to use these wire junctions because they let you attach only a single wire to each terminal screw (color-coded screws are connected internally within the wire junction). It is not a good idea to try to attach any more than two wires to a terminal screw. The screws usually are not long enough to handle the greater thickness of more wires and still tighten securely. Even a slightly loose connection at a terminal screw can cause a problem in getting the right voltage and current to the phone, although the connection may look solid to you.

PRELIMINARY CHECKS FOR WIRING

Before you finalize your plans for a telephone system to be installed in the wall or along baseboards, you must make some preliminary checks.

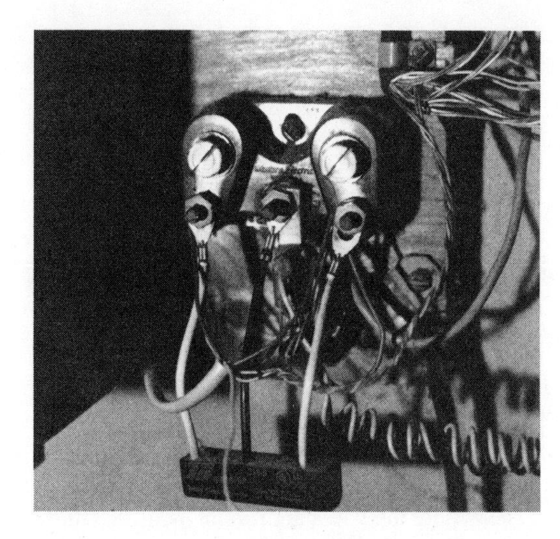

The large bundle of wires leaving this phone company protector feeds each individual outlet in this home. This is typical of the connections at a junction in cluster wiring.

Locate Electrical Wiring

Check where the electrical wiring goes in your home. If you have a set of the builder's plans, these wiring locations will be marked. If you're working only from a set of real estate floor

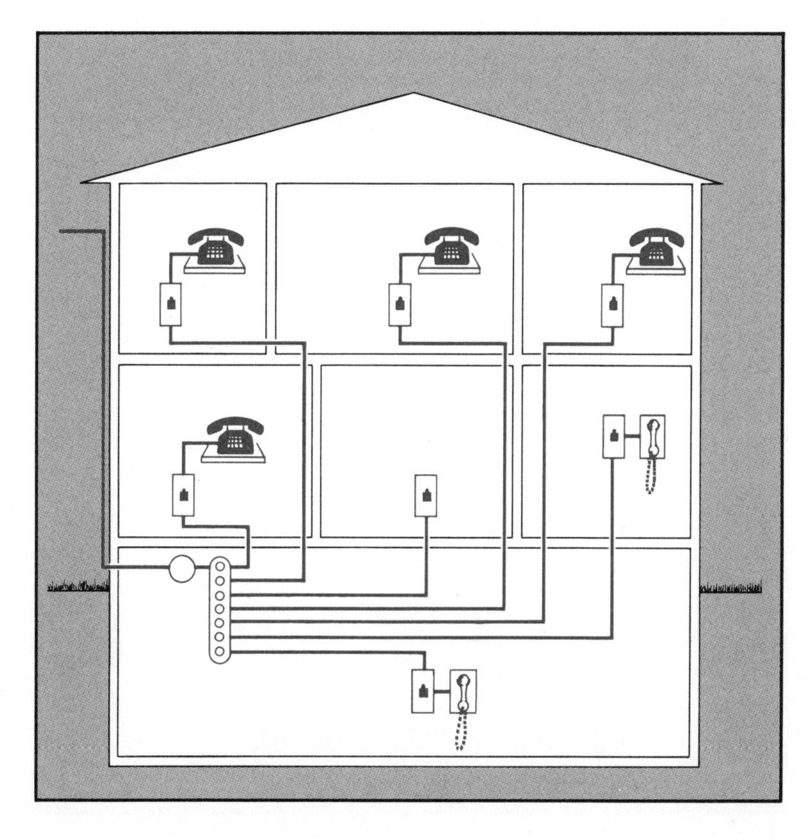

Cluster wiring, shown here, runs separate wires to each phone location from one central junction source. Many times phone locations in one-story homes are more easily reached from a basement or crawlspace with cluster wiring.

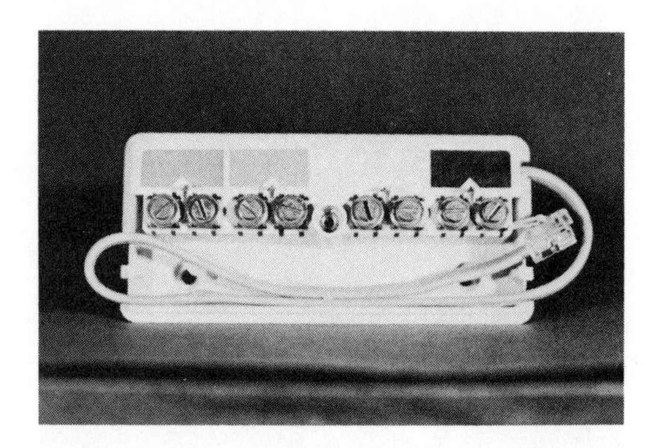

A wire junction lets you connect to several different locations from a single feed wire.

plans or from your own drawing, you'll have to guess where the wires are run by locating where the outlets, switches and fixtures are. It is safe to assume that if you have an AC outlet on the wall where you want to install in-the-wall wiring for a phone, somewhere in that wall are the power lines you'll have to avoid.

The electrical wiring can be run to an outlet or switch through the center of the wall (approximately at switch height) and then dropped down for wall outlets or run up for wall light fixtures. However, other builders will run the electrical feeds above the walls in the ceiling joists or below the wall through the floor joists and enter the walls for a vertical run at the location of the outlet, fixture or switch. Many times two wall outlets on the same wall will have the AC wire run directly between them, about 14 in. up from floor level.

You can safely plan to miss most electrical wiring if you plan your telephone wires between stud locations in the wall, low to the floor. Wall outlets are normally located about 14 in. from the floor. This usually gives you plenty of room below the outlet and above the sole plate (bottom 2×4 of a wall) to run your telephone wires. But be careful. It's entirely possible that some builders and perhaps a previous owner's remodel ran the electrical wiring along that bottom 2×4 in the wall.

Sometimes you can see where the electrical wiring is run to and from an outlet box by removing the cover plate and looking through the small space between the outlet box and the wall board. You can cut a little of the wall board away to see it better (about ¼ in. so the face plate still covers the opening). Often just where the wires enter the AC box will give you the information you need.

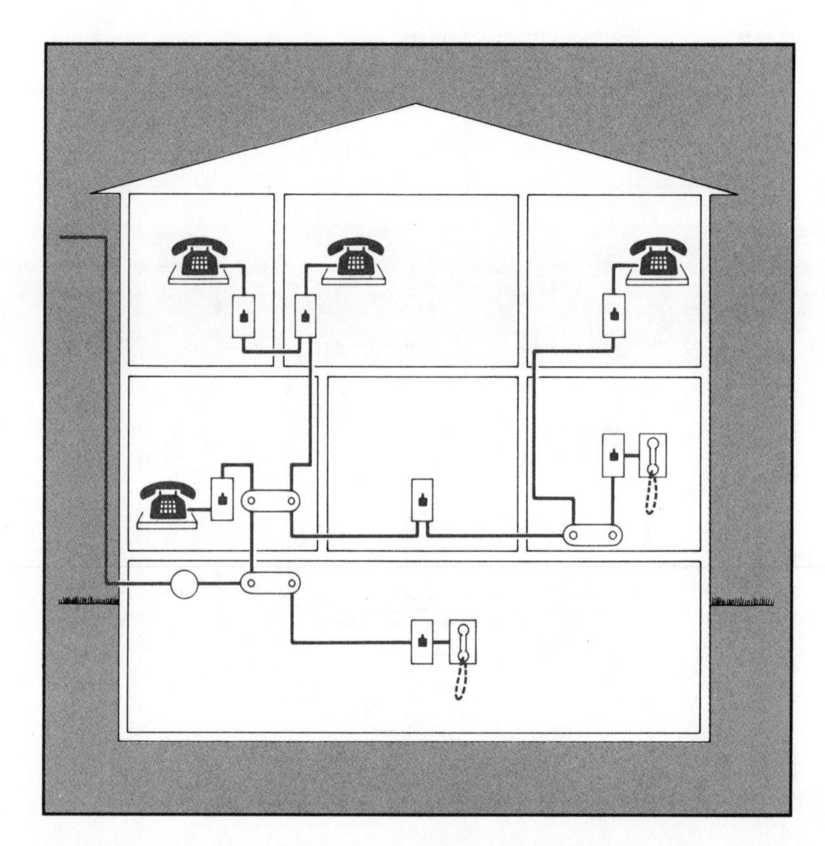

A combination of chain and cluster wiring will be used in most home installations. Wires are run in a chain to "hub" locations, where a junction provides cluster wiring branches to several nearby phone outlets.

Check Access to the Walls

Check access to each room from the area below and above the room itself. Many times the entire first floor of a home is accessible through an unfinished basement or crawl space. Even in some finished basements, you can get at the rooms above through removable ceiling panels.

When you are planning to run wires through floor joists in a basement or crawl space, you may have to measure from an outside wall to locate where the interior walls are located. In most cases, these walls are hidden from view by the subflooring. Some builders will run the electrical wiring through the floor joists below and up through the wall to outlets, fixtures and switches. This makes it easy to locate both the walls and the electrical wiring you want to avoid.

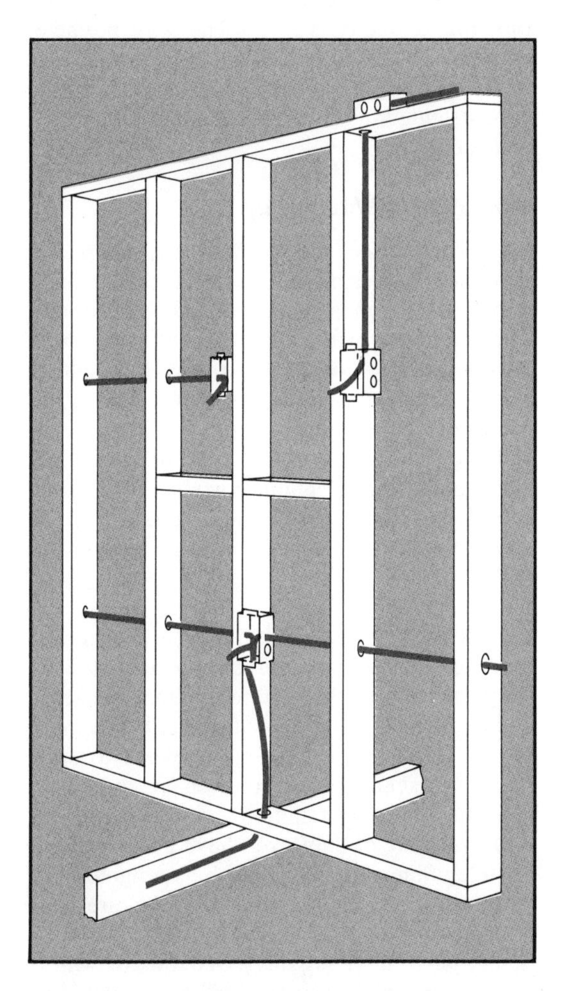

AC power lines in the walls must be avoided when planning hidden wiring. Builders or previous owners could have put the AC line feed in different places.

In second floor rooms, the walls are often accessible from an attic. You may have to remove some of the attic insulation or get it out of the way to find the wall locations. This will not do any permanent damage to your climate control if you're careful about putting the insulation back the way it was.

Often the top plate of a studded wall is more easily seen from an attic because the ceiling material does not rest on top of it.

If you can reach the room from below or from above, reflect this in the wiring scheme on your plans.

Check Room-to-Room Access

If you're planning a baseboard installation, make sure there are no AC electrical feeds or plumbing pipes in the wall where you'll drill through to reach the next room.

New Construction or Remodeling

If you are building a new home or doing a major remodeling that includes new walls in which you will want to install phone outlets, plan for these outlets before construction is finished and wallboard is put up. Often you can locate interim studs at the exact location you'll want to put a wall phone, making future installation both easier and more substantial than if you

Seeing where wires enter an AC wall outlet will give you a clue about where wires are located in the wall.

Spaces inside the walls between rooms may be reached through the floors from an unfinished basement or crawlspace.

In some finished basements you can get at the rooms above through removable ceiling panels.

tried to mount a wall phone with expansion bolts on only the wallboard itself.

MAKE A PURCHASE LIST

You can now make a detailed shopping list for the supplies you'll need for the actual installation.

How Much Wire Do You Need?

From your scale floor plans, measure the distance between each outlet you plan to install. Measure where the wire will go, not just the straight-line distance between the outlets. For example, if your plans call for wiring stapled to the baseboard, you'll have to allow enough wire at each doorway to run up the door jamb, across the top molding, and down returning to the baseboard on the other side. This could add 12 feet of wire at each doorway you have to cross.

On the other hand, you may not have to run your wire to a doorway and back along the wall of the next room if you can simply drill a hole for the wire through the wall.

If your plans call for a wall phone installation, be sure you add the distance from where the wiring will be run through the walls up to the phone and then back down to continue the run to the next outlet.

Allow yourself at least 12 in. (24 is better) at each end of a wire at a terminal box or jack location. This gives you enough room to work when you strip the wires and attach them to the terminal screws.

Calculate your wire lengths to the nearest 25- or 50-foot length. If you are buying multiconductor telephone cable, you may not find places that will stock it in bulk and cut exact lengths for you. Most of the time it is sold in 50-foot rolls. Some stores stock 25- and 100-foot rolls. (In a well-equipped store, you may find a 500-foot roll, but this is rare.) The wire is cheap enough, usually about 10 cents a foot or less (watch for sales!), so being a little generous in your calculations is not going to be a major financial risk.

What Kind of Wire to Buy

Bulk Wire. You will need to buy telephone wire, not AC extension or lamp cord that's sold in bulk rolls at most hardware and home stores. The telephone wire must have at least three conductors, be solid not stranded, insulated, and of 22 or 24 gauge (larger AWG wire if the runs between outlets are long). It will make installation easier if you buy rolls of wire especially made for telephone installations. These are always the correct gauge and are color coded the way all other wiring in the system will be.

Color-Coded Wire. The red–yellow–green–black coding is now universally accepted as standard among the different phone systems and phone equipment manufacturers. You must have at least these color-coded wires to each phone. You can buy four- or even six-conductor cable for special installation requirements. With a four-conductor cable you have an extra wire for telephone accessories such as a light or intercom buzzer. If you have two separate telephone lines (two directory listings) into your home, plan on buying the six-conductor wire if each phone in your system will be equipped to handle both lines. You will need special two-line phones for this kind of installation.

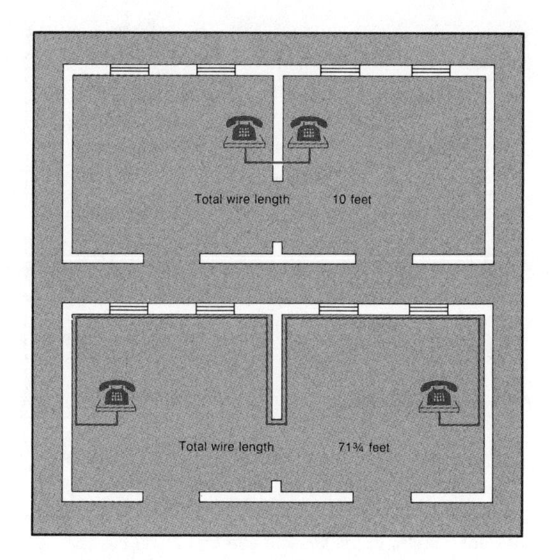

Running wires through a wall into the next room can save you many feet of wire and many hours of installation time.

Modular Extensions. If your home installation calls for using modular extension cords instead of new cable wiring, be sure the kind of extension you buy matches the jacks in both your phone and in your wall. There is a difference between cords. Telephone connection cords come with male plugs on each end. One plug goes into the wall jack, the other into the modular jack on the telephone itself. Extension cords, on the other hand, usually have a male plug on one end to plug into the wall jack and a female plug on the other end to receive the male plug on the cord from your telephone. Be sure you know your installation requirements before you buy. You can buy adapters to match the two male ends of mismatched cords, but with careful planning you won't need this extra piece of hardware — or the extra expense.

Be careful of extension cord lengths. Popular sizes include 6-, 10-, 15- and 25-foot lengths. If you're never going to move the phone off a desk or phone stand, buy one just long enough to reach from the jack to the desk without being unsightly along the wall. If you plan to wander around the room with the phone in your hand, you'll need a cord long enough to reach the outside limits of your planned excursions.

In some electronic and specialized phone stores you can buy the modular wire in bulk lengths and accessory plugs that you can put on the cords with a special tool — also sold separately in the same store. These may be more handy to have for repair of broken modular cords than for installing your phones, since the lengths of ready-made cords are varied enough to meet any need.

Add Junctions, Terminals and Jacks to Your List

From your drawing, count the number of outlets you'll need for your installation. You will also need wire junctions wherever you split your wire run to go to two different locations. If you plan to install wall phones, remember they take a different kind of modular connection plate than does a plug-in desk phone.

You'll find more help on selecting just the right modular hardware in Chapter 7 and specific directions on hooking them into your line in Chapter 12.

WIRING YOUR SYSTEM

After you've completed your shopping trip, you're ready to start actually installing your home telephone system. There are different installation requirements for a system using extensions from existing jacks, a system running wires along baseboards and moldings, and a system requiring in-the-wall wiring.

Before You Begin

Before you begin installing any wires or working with the modular connections, disconnect your system from the phone company at the incoming terminal. The 48 volts of direct current the phone company sends to your phone probably won't even be felt. You don't, however, want to be touching two bared wires if the 90- to 120-volt ringing current comes down the line because someone is trying to call you. Even at this voltage, the ring signal's current isn't likely to do most people serious harm, but the tingle or jolt can be quite a surprise when you least expect or want it.

Running Extensions

If you are expanding your telephone system by adding extensions to increase the flexibility of existing modular jacks, there are some installation hints and cautions you should know about.

First, always work with modular jacks and cords. If your home telephone wiring system still has the older four-prong jacks in or on the walls, now is the time to change them. Full directions to do this are in Chapter 12.

Remember to keep the extensions out of the way of possible damage. A telephone extension draped across a traffic pattern in a room is sure to be tripped over and permanently disconnected where the wires enter the plug. Extensions can be stapled to the baseboard or moldings to keep them out of harm's way. For these semipermanent installations, use only insulated staples and be sure you can remove the staples easily without gouging the insulation on the wire if you want to change the phone's location later.

Installing Baseboard Wiring

Fastening to Baseboards and Trim. Installing wiring along baseboards and molding requires that the cable be stapled to the decorative wood trim pieces. When fastening telephone cable to any surface, be sure to use insulated staples.

Your first inclination may be to staple the wiring along the top of the baseboard, right at the junction with the wall. This is usually where the cable is hidden the best. You don't, however, have a good enough base to hold staples here, either on the wall or on the very thin part of the wooden baseboard piece.

Another temptation is to nail the wire in the joint where the quarter-round and the baseboard meet or between the baseboard and the flooring. This is going to create the problem of nailing the staples with one tang into each piece of wood. More often than not, these have different densities, and you are doomed to bend more staples than you successfully install.

The best place to staple your telephone cable is at the flattest possible place on the baseboard where the wire can still be as inconspicuous as possible.

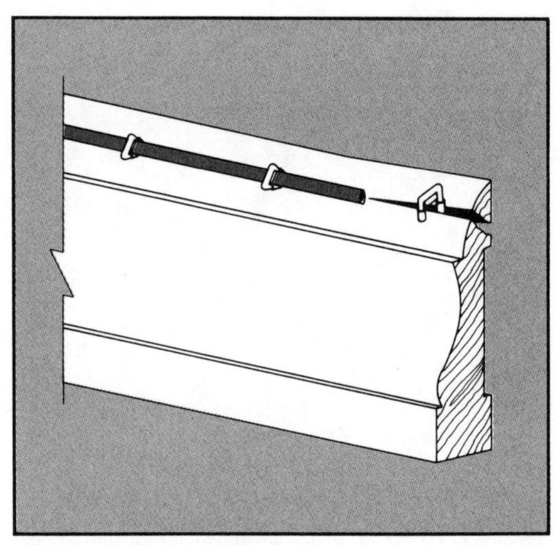

Staples may split the wood if you try to hide baseboard wiring where the wood is too thin.

Staples are almost sure to bend when you nail them with one point into hardwood and the other into softwood.

Changing Wire Direction. When you make the transition between the horizontal baseboard and the vertical run of a door jamb or other molding, leave a slight loop in the cable rather than using the last staple for a sharp right-angle bend. The wires in a telephone cable are solid pieces of copper, not stranded cable designed to be flexible. The solid-wire cable won't take a sharp bend without breaking the conductors.

Painting Surfaces First. Paint the baseboards and moldings before you install your wiring. Even today's latex paints will not do wiring insulation much good. Mysterious problems with telephones can often be traced to insulation breakdown within a cable caused by the chemical action of the solvents in paint on the plastic insulation material on the wires.

Drilling through Walls. You can run the wiring along the baseboards, then through the wall into the next room. This requires that you drill a hole large enough to feed the wire through. Don't drill the hole until you double check your plans and the actual wall location itself to be sure there are no electrical wires or water pipes hidden behind where you want to drill.

When you run the wire through the end of a wall, you will be drilling through about 5 in. of wood and plasterboard. There are almost always additional studs at the end of a wall to provide adequate support and a nailing surface

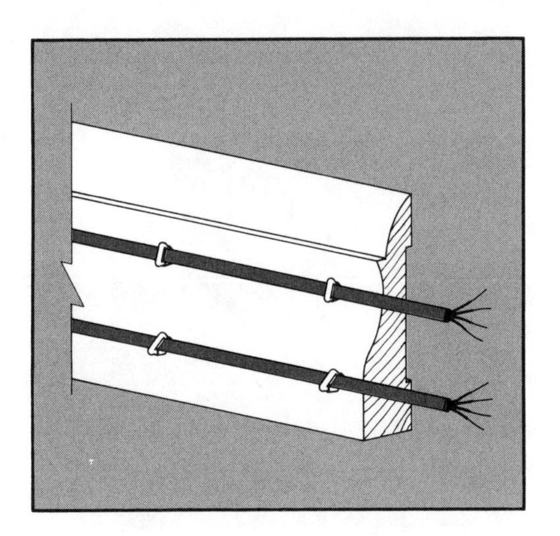

Wires can be hidden at several places by stapling them at natural grooves or projections in the baseboard trim.

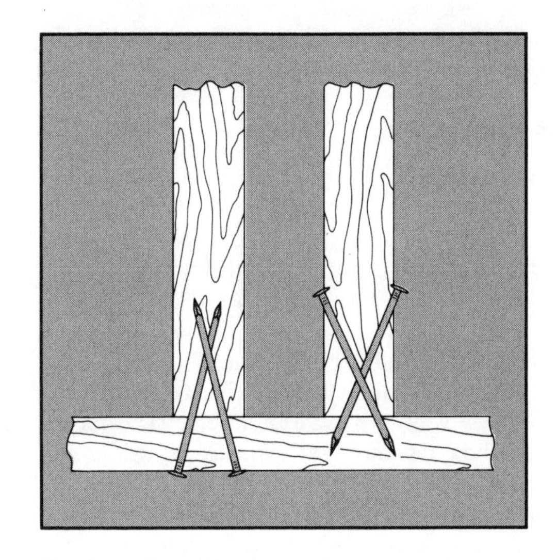

To miss nails inside your walls at the top or bottom when you drill a hole between rooms, be sure to drill at least 5 in. up from the floor.

to hold the two walls together. Nails from the bottom 2 × 4 plate will normally extend up into the vertical 2 × 4 studs about 1 to 1¼ in. The vertical 2 × 4 stud can also be toenailed into the bottom plate with nail heads about 2 in. from the bottom. To be sure you miss the nails with your drill bit, plan your hole to be at least 4½ to 5 in. up from the floor.

If you are passing the wire from one room to another in the middle of a wall, between the wall studs, usually all that stands in your way are two pieces of ½-in. wallboard. A long ¼-in. spade bit or a ¼-in. auger bit chucked in a hand brace or power drill can easily go through the wallboards — sometimes too easily. Go slowly! A helper is essential here. Have your helper look for the tip of the drill bit as it begins to dimple the surface or just begins to penetrate the other side. Stop drilling from that side and go into the next room to finish the hole from there. For almost all telephone wire you'll be running, you can get by with a ¼-in. hole.

Fishing the Wire through. It is not apparent how limp and unresponsive a short piece of telephone cable is until you try to shove it through a hole in the wall. You will save hours of frustration by using a stiff fish-wire pushed through the holes. Loop the telephone cable securely to the fish-wire. Pull the fish-wire back through the holes slowly and the cable will follow. While you can buy fish-wire especially for this purpose, for just going through a standard thickness wall you can use a straightened coat hanger. Most of these already have handy hooks bent in them for the wire.

Attaching the Hardware. You can attach your junction boxes and modular jacks to the baseboards using regular roundhead wood screws. These almost always come with the telephone hardware. While you can use the coarser threaded plasterboard screws to fasten boxes directly to the wallboard, you will be well advised to use plastic sleeve anchors in the wall. Unless you're installing a mounting plate for a wall phone, you don't need the greater load capacity of a toggle or Molly bolt. The plastic sleeve or similar wallboard anchor is sufficient to take the stress put on the wall connection

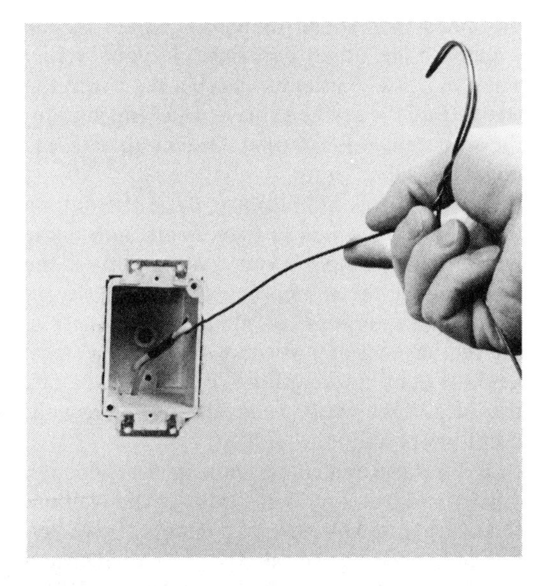

A wire coat hanger can be used as a fish-wire for short jobs. Secure the cable to the built-in hook.

from plugging the jacks from your phone extensions in and out.

In-the-Wall Wiring

Hiding the wires for your telephone installation requires that you have access to the spaces inside the walls. You also have to plan for some adequate way to terminate the wiring for the modular jack or terminal box.

Are Receptacle Boxes Necessary? While local codes usually don't require that telephone cables in a wall terminate in an electrical receptacle box (as most AC wiring must), it still makes good sense to plan on using them. All telephone wall jacks and other wall-mounted hardware come with a wall plate that has mounting holes spaced to fit the screw lugs on a standard electrical receptacle box. These wall plates are the same size (and usually the same color) as the plates used to cover switches and electrical outlets. This lets you apply the same principles of installation you would for electrical wiring.

Installing an Electrical Receptacle Box. Most electrical boxes sold in hardware and home stores are designed to be put in the wall before the wall covering is installed. These usually have holes in both sides of the box in which long (10d) nails fasten the box to the studs.

You can buy boxes specifically designed for installing in walls that already have the wallboard put up. These boxes have spring clips on the sides, adjustable to the thickness of the wallboard or other covering on your walls. They're a little harder to find, but it's worth the effort. Buy these when you're installing your in-the-wall telephone wiring system in existing covered walls.

To install this kind of box, trace the outline of the box on the wall surface. Some units come with a paper template you can tape up on the wall to make the appropriate cuts. Drill two entrance holes for your saw at opposite corners of the rectangle drawn on the wall. Using a sharp keyhole or plasterboard saw, carefully cut along the lines. If the waste piece falls inside the wall, don't worry about it.

Before you install the box in the opening, you should fish your wires through the wall out the opening and into the box. This will give you one less place to fish around with them when you have to grope inside finished walls. When you're ready, the box snaps into the opening

This receptacle box is designed to install in walls that already have the wallboard in place.

Snap this type of box into a hole cut in the wallboard after the wires have been fished through. Snaps on the box hold it snugly in place against the wallboard.

you've cut and the spring latches snap behind the wallboard, holding the box securely. This type of outlet box is strong enough, in most cases, to handle the weight and stress of a wall phone and its mounting plate.

Installation without Receptacle Boxes. All the telephone hardware you will need can also be purchased for flush mounting, without having an electrical receptacle box recessed into the wall. These units simply screw into the wallboard or moldings at the places where you retrieve the wires from inside the walls. These units usually cannot be anchored as securely as one attached to a receptacle box, but they are more easily moved to a new location along the wall without extensive wallboard patching projects. The advantage of having a smaller hole in the wall through which the wires enter the telephone box becomes a disadvantage when you try to fish wires through the smaller opening.

New Construction. If your installation is being made before the wall covering is installed, the task of running your telephone wires is simply a matter of locating an electrical junction box where you want the modular jack or wall phone and running your cable through ½-in. holes in the stud to the box.

Don't fasten these wires to the studding inside the walls. After the wallboard is installed and the walls finished, you may need to replace a three-conductor cable with a six-conductor if you add a second line in the house. If the wires are not fastened, the old wire can be used — usually successfully — as a fish-wire to draw the new cable through the walls, replacing the old as it is removed. That's also why at least ½-in. holes through the studs are recommended even though most telephone cable would comfortably fit in a smaller hole.

Don't run your telephone wiring in the same conduits or openings cut or drilled for your AC wiring. Steer clear of water pipes, too. Keep the phone wire as far away as is economically possible.

If you're making this installation in a new home, be sure to tell the builder what you're doing. You don't want to find out after the wall covering is in place that some carpenter or electrician who didn't know what it was removed your wiring. Maybe, if you ask nicely, they'll do the wiring for you!

Entry into Existing Walls. The easiest access to the inside of the existing finished walls is from the areas above or below the room.

If you have access to a room from the area below, either through a crawlspace or from an unfinished basement, you'll need to run the wires along the floor joists and then up into the insides of the wall.

When you run the wires to the location where you plan to enter the wall, be sure to attach the wires (loosely) on the side of the joists, not on the exposed bottom edge. Any wiring there, although it is tempting to put it there when you have to cross ten joists to reach the other wall, will inevitably be in the way later when you want to install a ceiling in the basement room or insulation in the crawlspace.

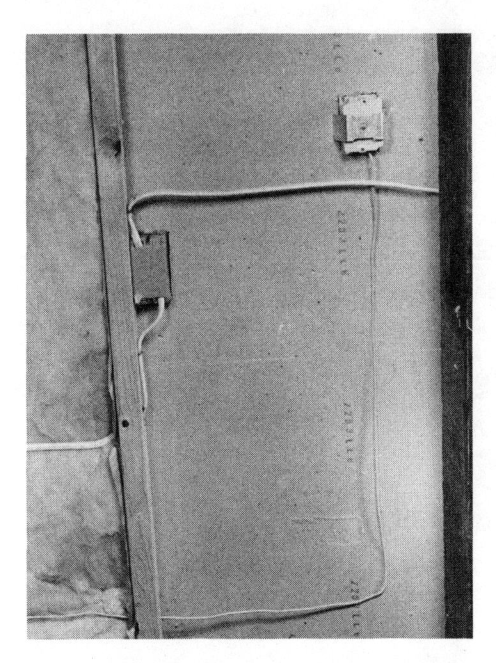

Telephone wires inside the walls should be kept well away from AC wiring. Cross AC wires at right angles to reduce hum and noise picked up in your phone lines.

Where you have to reach a location requiring that the wires be run across the joists, drill a ½-in. hole through each joist and string the wire through. If your joists are on 16-in. centers, you may not be able to fit an electric drill with a long spade bit in the space between them. In these cases, drill holes at an angle from the open area where you can work.

Run the wire to the location of the wall you are going to enter to install your telephone outlet. When you've located the wall from below (and have double-checked both your plans and the wall itself to be sure there are no electrical wires or water pipes in the way) drill a ½-in. hole up through the subflooring from the basement or crawlspace, through the bottom 2×4 and into the wall. Be sure your measurements are correct! You don't want to cope with repairing a floor because you missed the inside of the wall.

If your outlet is to be located on an outside wall, you won't be able to drill straight up from the basement or crawlspace into the wall. The bottom 2×4 of these outside walls almost always sets directly over the foundation, or over the top plates of the wall below. For this installation, you'll have to drill a diagonal hole up from the basement or crawlspace through the subflooring and bottom 2×4 inside the wall.

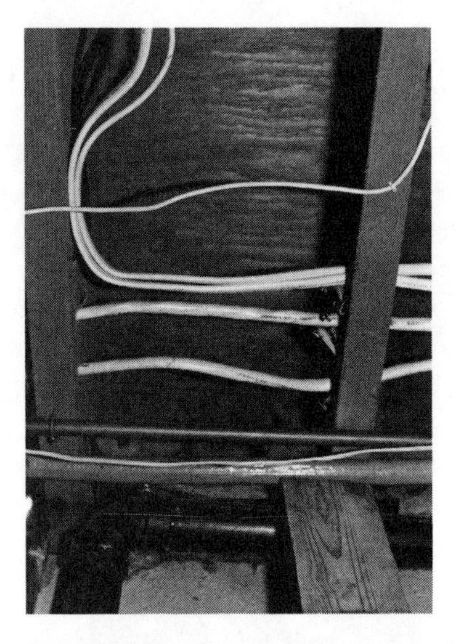

Run wire along the sides of floor joists. Wiring stapled to the bottom edge will get in the way if you want to put a ceiling up later.

Once you have successfully penetrated the inside of the wall, run a fish-wire from below up through the hole you drilled. You should see that fish-wire through the opening you've cut in the wall covering for the outlet box. If you can reach it, pull it through the opening. If you can see it but can't reach it, you'll need another fish-wire to hook it and bring it through the opening where you are going to start the cable. Be sure the other end is still sticking out of the opening in the wall. You need the hooked end of the fish-wire to fasten your cable when you pull it through the other opening.

If you're going into the walls from above the room, the process shown here for coping with basement entry can just be turned upside down.

Handling Special Problem Walls

When you try to fish wire through a wall, it is not uncommon to encounter a 2×4 nailed between the studs about halfway up the wall. These cross studs serve as additional bracing for the wall and are sometimes called *fire blocks* because they retard the spread of flames within the walls in case of fire.

You can, of course, rip out an entire section of wallboard if you encounter this obstacle. Then you have free access to the wall — and the rather unpleasant and relatively expensive project of replacing the wall and refinishing the entire room. A more economical answer is to cut a small slit in the wall for the wire, and then patch it with joint compound.

You first have to locate the offending piece of 2×4. Use a stud locator or thump on the

After you carefully measure where the wall above is, drill through the floor into the space between the walls.

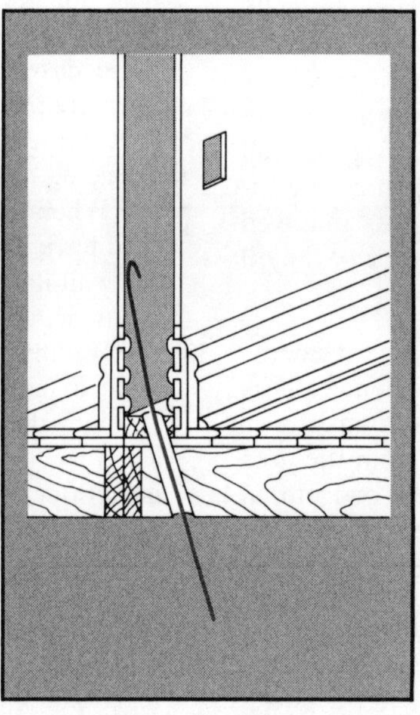

The first fish-wire is run into the wall from below through the hole drilled in the floor to the wall.

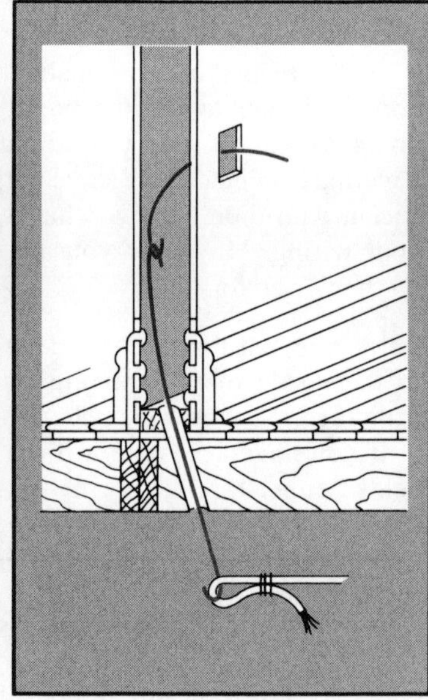

Then carefully pull the fish-wires with the cable attached through from the outlet location in the room.

wall. Sometimes it takes a trained ear to tell the difference between the solid sound at the studs and the hollow sound between them. A better way is to run a fish-wire through your access hole until it is stopped by the wood. Remove the fish-wire and measure how far it went into the wall. Then, from the floor or ceiling you can locate the obstacle (after subtracting the thickness of wood you drilled through beneath the floor or over the ceiling).

Drill a hole through the wallboard directly below where you (hopefully) have determined this piece of wood to be. With a utility knife or a small keyhole saw, cut a narrow slot in the wallboard across the 2×4. This opening does not have to extend more than 1 in. in either direction above and below the 2×4 and can be about ¼ to ½ in. wide.

Clear the wallboard away in the slot, exposing the wood behind it. Then, with a sharp chisel, make a ¼- or ½-in. slot in the wood, about halfway into the 2×4.

Use the fish-wire technique from above and below to get the telephone cable through the openings above and below the 2×4. After the wire is successfully run to its final location on the wall, slip a piece of sheet metal between the wallboard and the 2×4 to protect the wire in the

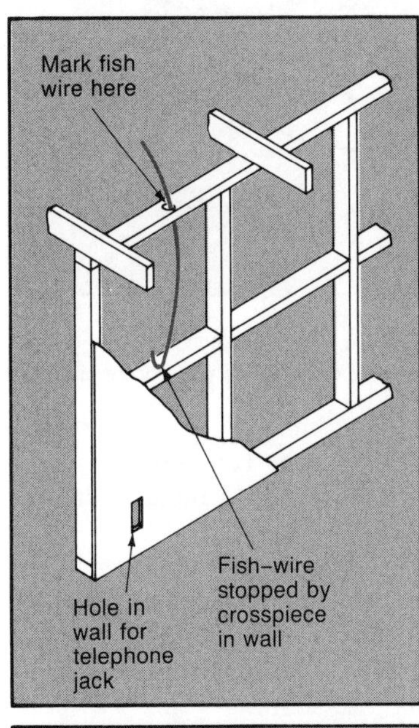

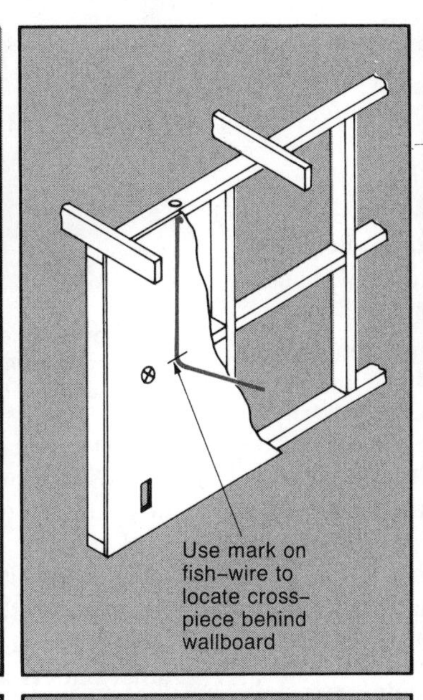

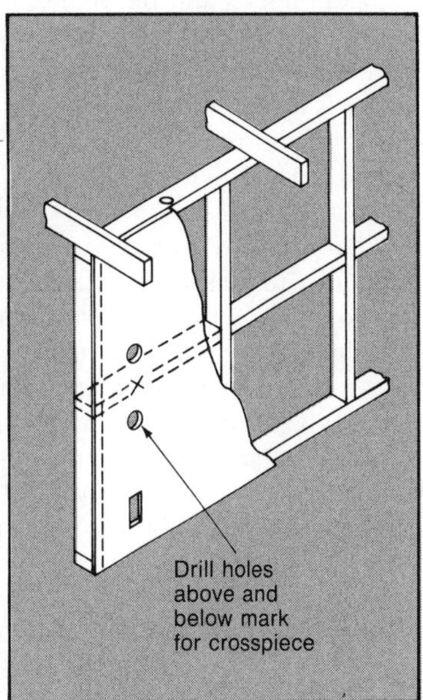

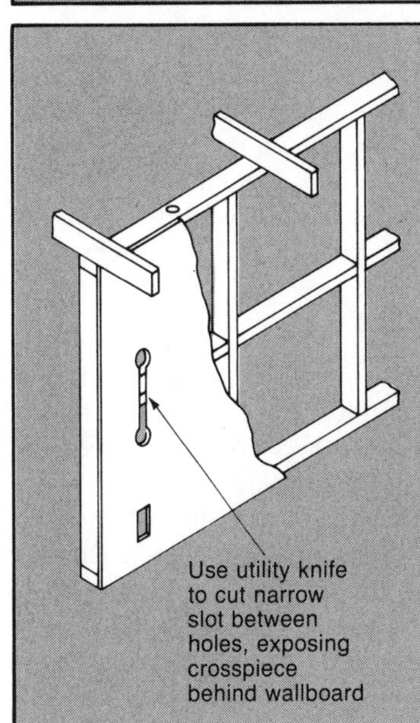

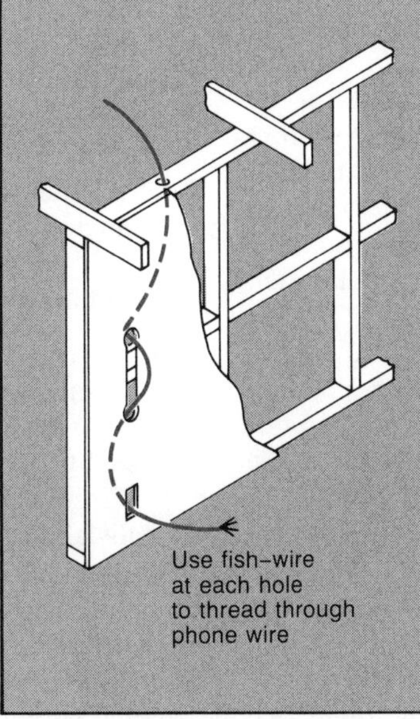

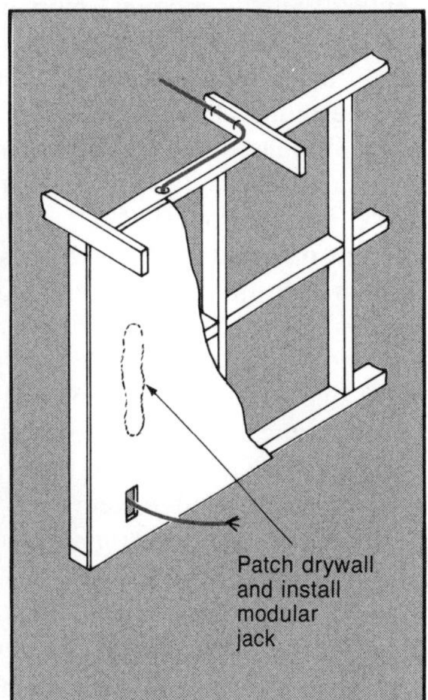

slot. You can also use a wood shim that you cut to be flush with the edge of the 2×4 but still give the cable at least ½ in. of clearance for movement.

All that's left to do now is to patch the narrow slot you've cut in the wall covering. In most cases, just several coats of joint compound will do the trick. Let the compound dry and shrink between applications. If you've made the slot in the wallboard too wide, you may have to feather the edges of the opening with a utility knife, and use joint tape with the compound to seal the slot. After the compound dries, wipe it gently with a moist rag to smooth it, and then repaint the wall.

If you are committed to an in-the-wall installation of your telephone wires in a home where there is no access to rooms below or above, you will have to use this technique every 16 or 24 in. to cross individual wall studs as you run the wire to its next location.

Another way to run wires in these homes is carefully to remove the baseboards. Then cut a narrow slot in the wallboard about halfway up the height of your baseboard material. Install your wires in the slot and replace the baseboard, taking care not to nail into the slot where you have the wires. If you opt for this technique, you will most likely want to mount your jacks and terminal blocks along the baseboards over a hole drilled to run the wires through.

Special Wiring for Computer Data Transfer

If you communicate frequently with other computers with a modem, having an extension phone picked up in your system can seriously disturb the transfer of data. You can wire the phone outlet at your computer with a special switch to disconnect all others in the system.

For this installation, you'll need a double-pole, single-throw (DPST) switch and a box to mount it in (although there is usually room for the switch on the wall-mounted modular jack plates).

The "privacy phone" outlet should be the first one in the chain from the telephone company's junction. You may have to adjust your wiring plan if this location is not the closest one to the start of your system.

Run your wires to this outlet in the normal way. Instead of going directly to the next phone (and all the others) in the chain, however, run the red and green wires to two poles of the DPST switch. The yellow (ground) wire can be connected as usual. Then, from the other two

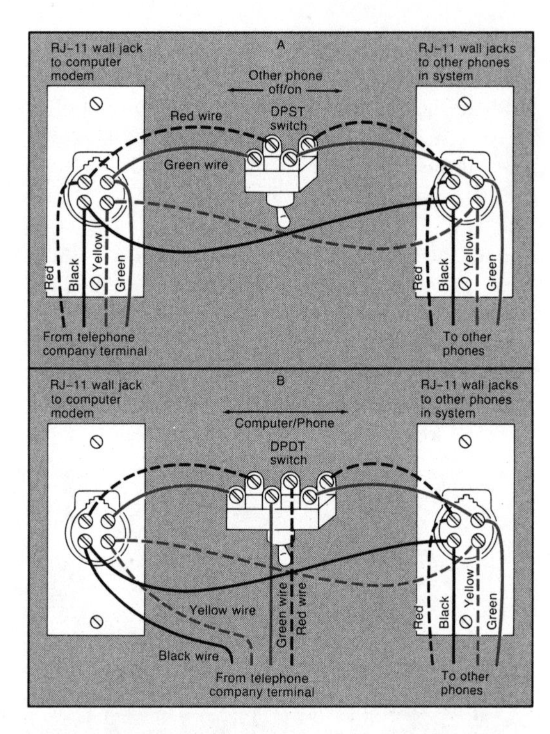

Connect the phone wires to the switch, following the color-coded connections carefully so you don't switch the red and green wires.

terminals on the switch, run red and green wires, along with the yellow ground to the rest of your system.

When the switch contacts are open, your telephone system ends there and does not reach the rest of the outlets. When the switch is closed, the signal is sent along the wires to all other outlets in the system.

While this system will protect your data transfer when you're communicating by modem, it also will cut off all other phones in the house as long as the switch is open. You might want to install large signs in your computer area to remind you to close the switch when you're done until the habit becomes firmly ingrained.

Running the Wires

When you run the wires to each jack or terminal location, be sure to leave at least 12 or 24 in. of wire to have something to work with when you install the jacks and connectors.

Fishing long lengths of wire through multiple walls almost always calls for a helper. Doing it by yourself will almost certainly cause you to end up with a hopelessly tangled mess of wire that has come unrolled from the coil or spool and clogs up your carefully planned and executed wiring route.

12
Hooking Up the Hookups

In the previous chapter you learned how to plan and install the internal wiring needed to put wall outlets throughout your home. This chapter will concentrate on the parts and materials you will need to connect the wiring in your system to the phones themselves.

The Modular System

The modular system was adopted by the phone companies to make installation easier — both for you and for the phone company service technicians.

Never Splice or Solder Wires. There is one overriding rule you must follow in connecting the final terminations to your phone system. Never splice or solder any wires in the phone system. There are enough connectors, terminals and other wiring aids for any do-it-yourself project to make this rule an easy one to follow.

Modular Plugs and Connectors. The modular connector system is built around two tiny basic units. One is a female connector, the other a male plug. Each of these contains four copper wires, laid out in a close parallel pattern. When the plug is inserted into the connector, the two sets of four wires come together in a tight grip. The four wires are *tip, ring, ground* and an extra wire that is not used.

The "official" code for these familiar home fittings is RJ-11. There are related RJ series fittings for special installations, especially in offices and large commercial systems.

The modular plug is designed so it can only be inserted into the connector one way. This means the four wires in the plug are always in the right position to touch the four wires in the jack. The tip wire in the male plug touches the tip wire in the female connector, the ground wire in the plug touches the ground wire in the connector, and so on.

Before You Begin

Before you begin installing any wires or working with the modular connections, for your own protection, remember to disconnect your system from the phone company at their incoming terminal.

Types of Hookups

You'll need modular jacks, wire junctions and perhaps some adapters or extensions to fine-tune your wiring system to the equipment you have.

Wiring Junctions. With the shift to the modular system by the telephone companies other pieces of hardware became available to help the do-it-yourselfer, making wiring a phone system easier. One of these is the wiring junction.

In the modular version of the wiring junction, the incoming modular plug is inserted in one of the terminals and the outbound jacks

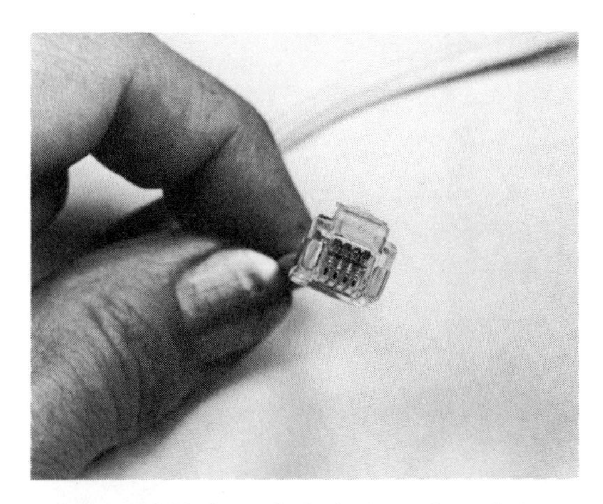

Inside an RJ-11 plug or jack, the four color-coded wires are attached to tiny metal strips that make the connections.

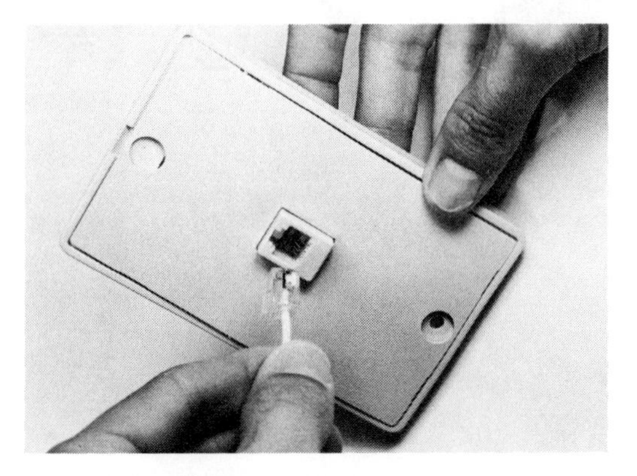

All modular jacks and plugs are keyed so they can only be inserted one way, ensuring correct matching of the color-coded wires.

This modular wire junction gets the signal from a modular jack or a network interface and has five outlets for phones or extensions.

serve the several extensions from the terminal junction.

There is also a screw-type junction for a wiring scheme that uses bulk-type telephone wire without modular connectors. The principle is the same. The incoming triad of wires is screwed to one set of terminals and the outbound wires are fastened to the other sets of screw terminals. On some, the inbound wires are from a modular jack and plug. These are connected internally within the junction to the terminal screws. The only precaution you have to take in installing one of these is to keep the color coding straight.

Think of a junction as a Y-adapter for your permanent wiring. Use junctions when you need to branch the wires to two locations. If you need to go to more than two, you may have to chain two or more junctions together.

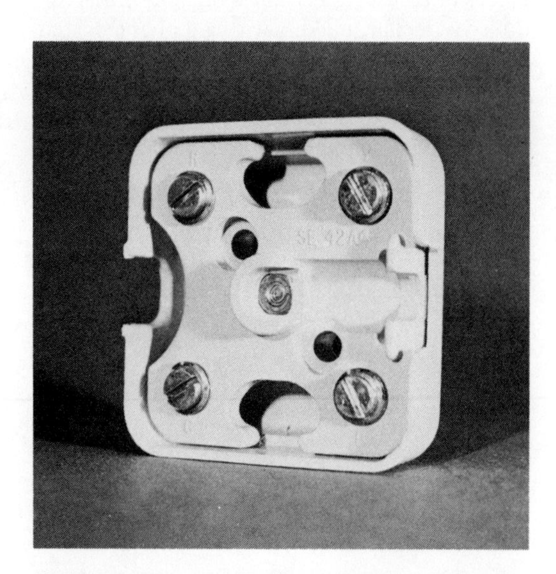

Each set of screw terminals on this junction is marked to keep the proper color-coding of your wires for the next location in the system.

Wall and Surface Mount Hardware. Almost all the hardware you'll need to finish the installation of your telephone system can be purchased for either mounting on the surface of a wall when you run wires along the molding, or recessed into the wall for flush mounting when you run wires inside the wall.

Mounting Surfaces. Surface-mounted modular jacks and terminals need to be fastened securely enough to let you plug and unplug the wires without loosening the piece from the surface. Almost all surface-mounted telephone hardware will come with attaching screws. These are generally good enough to fasten the piece into the solid surface of a baseboard or other wooden molding. If you're mounting the jack or terminal directly to wallboard, you may want to substitute the coarser threaded wallboard screws or use the plastic sleeves that insert into a hole you drill in the wallboard.

For flush-mounted hardware inside the walls, it is much easier to use an electrical junction box (provided you planned for this in your wiring scheme). The plate-attaching screw locations on this telephone hardware match the mounting lugs on these standard electrical boxes. It is also possible, however, just to cut a hole in the surface large enough to accept the hardware mounted on the back of the plate and

A wall plate modular jack is used where wiring is hidden in the walls. Attach it the way you put on an AC switch or outlet cover.

then fasten the plate to the wall. Most wall-mounted hardware will come with appropriate fasteners for this kind of installation.

You can buy special wall boxes now for telephone installations. These are made of plastic (to provide insulation for wire connections) and have screw lugs inside to mount the hardware or wires.

Where the Wires Should Go

To ensure both correct and safe operation of your telephone system be sure to use wire that lets you follow the standard telephone color coding. At each connection you make, always be sure that red is attached to red; green to green; yellow to yellow; and black to black.

Connections to the Phone Company's Terminal. The first place to start installing hardware for your wiring is at the telephone company's terminal where their lines enter your house. To make any later changes easier and to minimize the risk of disconnecting or damaging the phone company's connections, it's wise to install a wire junction of your own at the phone company's terminal. Run the wires to the phone company's terminal, but don't hook them up until you've finished hooking up all the bare wires. You don't want to be surprised by ringing voltage coming down the line while you're working.

Carefully run the four wires (three if the phone company has not connected the black wire) to your wire terminal. This way, you can do all of your work from now on from your own

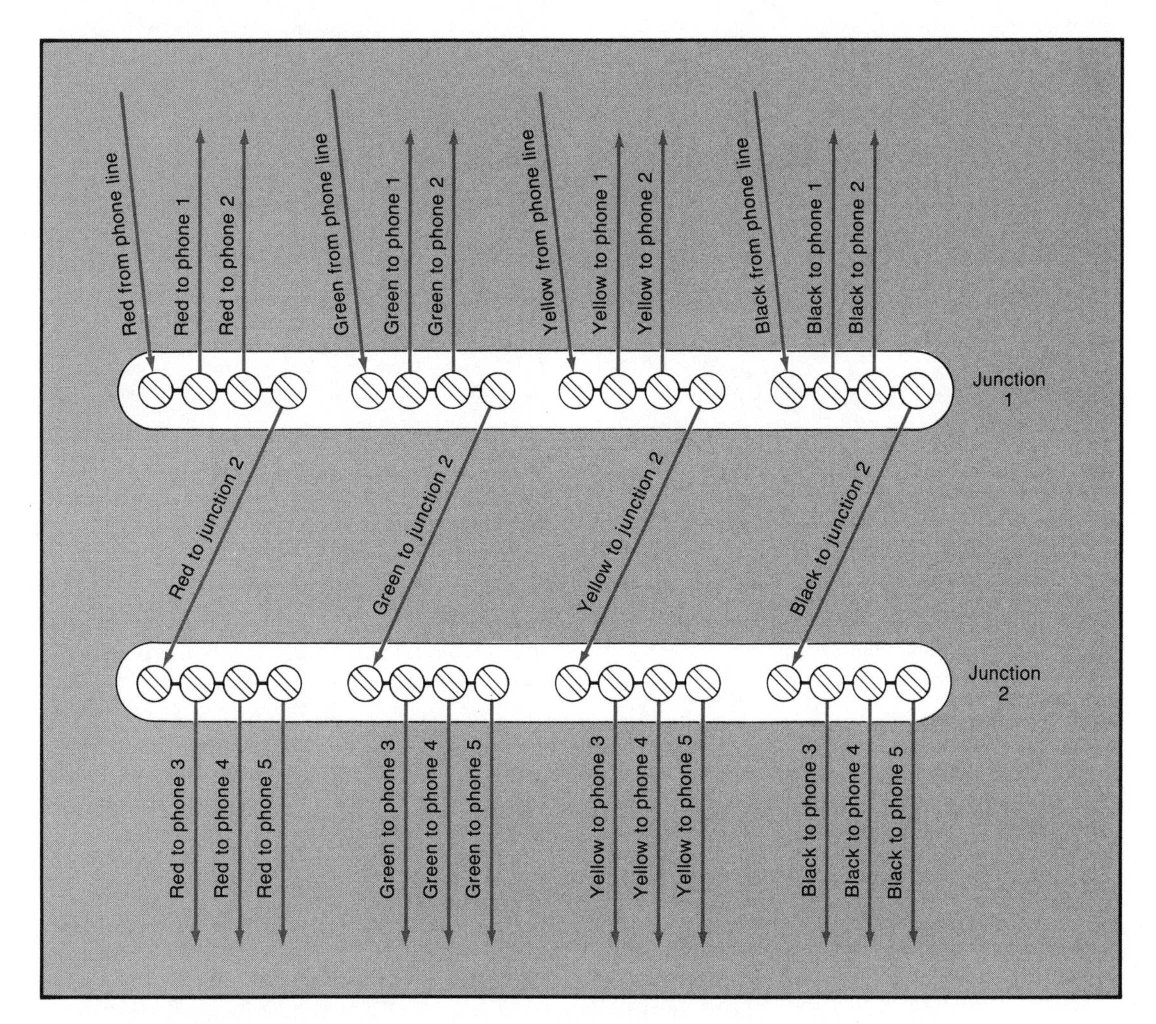

Modular or screw-type junctions that do not have enough outbound terminals to branch to the phones at that location in your system can be "chained" together by connecting the set of connections at one output to the input of another junction.

This kind of modular jack is used where wiring is stapled to the baseboard or trim molding. It attaches to the wall or baseboard.

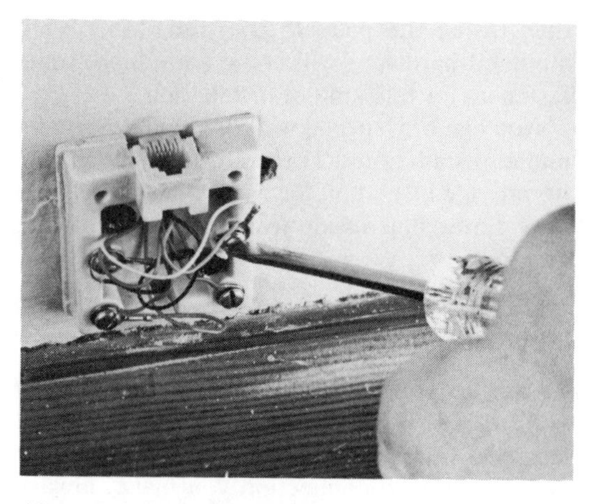

After you connect the wires on a surface-mount jack, screw it to the wall or molding and staple wires neatly in place.

You can install a flush-mounted modular jack over a standard receptacle box you use for AC wire installations. The wiring terminates inside the box.

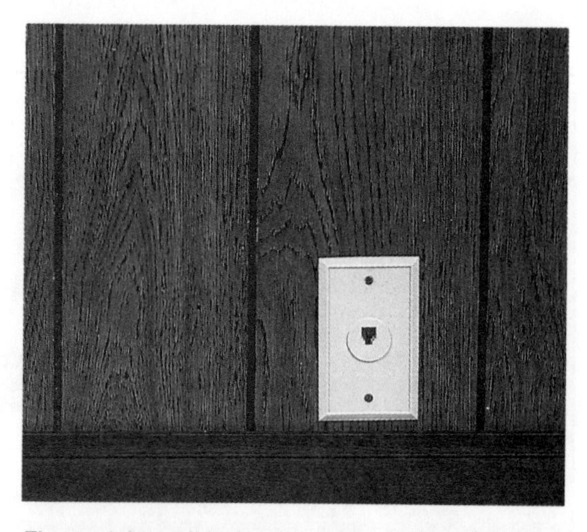

The modular wall jack can be screwed directly to the wallboard after cutting a hole just large enough to clear the screw terminal projection on the plate.

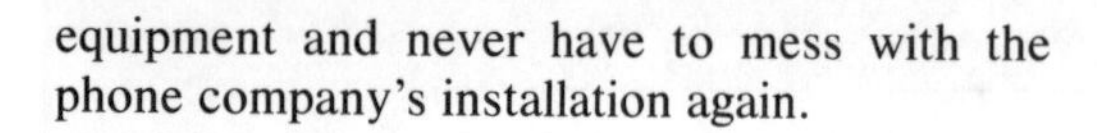

equipment and never have to mess with the phone company's installation again.

How to Strip a Wire

If you have been generous in the planning and running of the wire for your telephone system, you should have about 12 to 24 in. of wire to work with at each place where you'll make a connection to a jack, terminal or mounting plate. You will actually need only about 2 in. of wire to make the final connections to the terminal screws on the jack or other hardware.

Before you start cutting the wire to length, check to see how much of it can be shoved back into the wall opening after the connections are made. If your in-the-wall installation terminates at an electrical outlet box, you will probably be able to hide the entire extra length this way and you won't have to cut the wire at all. If you're working with surface-mounted hardware with wiring stapled along the baseboards or moldings, you are going to have to work with shorter wires.

To give you the extra room to work with baseboard wiring, remove the first couple of staples after you cut the wire to the correct length for a neat fit. You can staple it back to the baseboard after you attach the modular jack or wire terminal you are hooking into.

Exposing the Wires. The first step is to cut

away only as much of the outer insulation of the cable that you'll need to expose the length of wires you'll need to make the connection. You want to be sure that the outer insulation is intact where the cable enters the connector and that the four individual wires are exposed only in the connector.

To cut away this outer insulation, start at the end you've cut, and with a sharp knife or diagonal wire cutters slit the insulation material to the point you've marked for it to be removed. Be extremely careful that your knife blade or the wire cutter does not nick the insulation on any of the four wires inside the outer case.

An easy way to make the cut around the wire at your mark is to pull the four wires inside through the slit and out of the way of the scrap piece of outer insulation you're going to remove. Then trim the excess away with your knife and wire cutters, being very careful when you get close to the wires inside.

Exposing the Conductor. You now have to expose the copper wire inside each of the wires you're going to attach. If the black wire isn't hooked to anything, don't strip the end off it. Just twist it into a coil and store it inside the connector safely out of the way where it won't contact any of the terminal screws. This is neatly done by coiling the wire tightly around a pencil or small dowel.

With a diagonal wire cutter or sharp knife, cut through the red, yellow or green insulation about ⅝ in. from the end of the wire. Be very careful that you don't score the copper wire too deeply. It's almost impossible not to score the wire a little when using a wire cutter or knife to remove the insulation. If the score is too deep, however, the wire will be severely weakened there and will most likely break off when you try to bend it around the terminal screw. If you use a sharp modeler's knife or a single-edged razor blade, you can cut the insulation successfully by stopping when you feel the resistance of the harder wire. It's a little more difficult to "feel" the correct depth of cut with a diagonal wire cutter unless you've had a lot of practice to develop your touch.

Once this inner insulation has been cut, use your finger nails or the diagonal cutters held loosely on the wire to pull the insulation off the end, exposing the bare copper wire.

There is an electrician's tool that is specifically designed to strip the ends from wire without scoring the copper conductor. There are several V-shaped grooves or sized holes in this tool, each corresponding to a different size of AWG wire. As the handles are squeezed together, the insulation is gripped, cut and pulled off all in one motion. If you have one of these in your toolbox, by all means use it. It is not worth the expense of buying for just this installation job, however, unless your future plans include a lot of electrical work.

At some phone stores you can buy a wire stripper tool. This tool has a built-in measuring scale that lets you strip a multiconductor cable without damaging the wires. It is intended for the common AWG size of the telephone cables but you may find other uses for it after your telephone system is wired. These little tools are not expensive.

Attaching Wires to the Terminals

Almost all telephone hardware has either screw-type or knife-type terminals to which you attach the wires. If the ones you consider buying require soldered connections, you'd be well advised to look someplace else for equipment easier to install.

Attaching to Screw Terminals. With a needle-nosed pliers, hold the copper conductor so you can bend the wire into a clockwise U-shape. Loosen, but do not remove, the terminal screw to slip on the U-shaped loop of wire. Be sure the wire goes clockwise around the screw — the cut end on the right as you look down on it. Close the end of the loop slightly more with the needle-nosed pliers, then tighten the screw while pulling slightly on the wire to keep it seated against the threads.

If you have wrapped the wire correctly around the screw, as you tighten it by turning clockwise the end of the wire will tend to turn with the screw wrapping it more securely. If, on the other hand, you have wrapped the piece of wire counterclockwise around the screw, the loop in the wire will separate and tend to unwind from the screw as it is tightened.

If there are washers on the terminal screws, they go on top of the wires. The washers keep the wire from turning with the screw.

You may have to attach two wires to each terminal screw if you are running the wires to the next modular outlet. Follow the same procedures outlined above with the first wire. Then loosen the screw enough to wrap the second wire around it and secure that one in the same

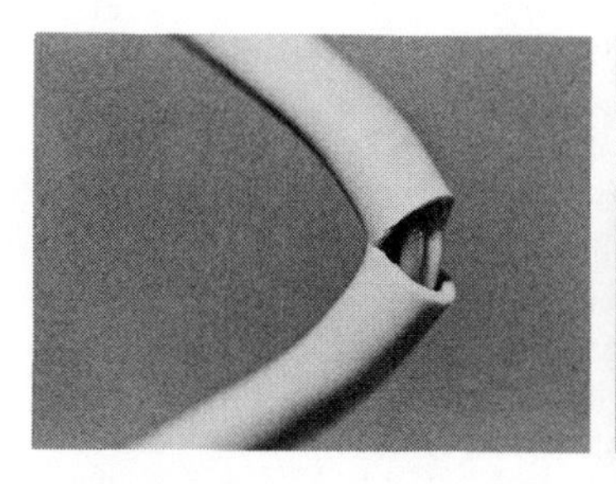

Carefully slit the outer insulation with a sharp knife, being careful not to damage the wires inside.

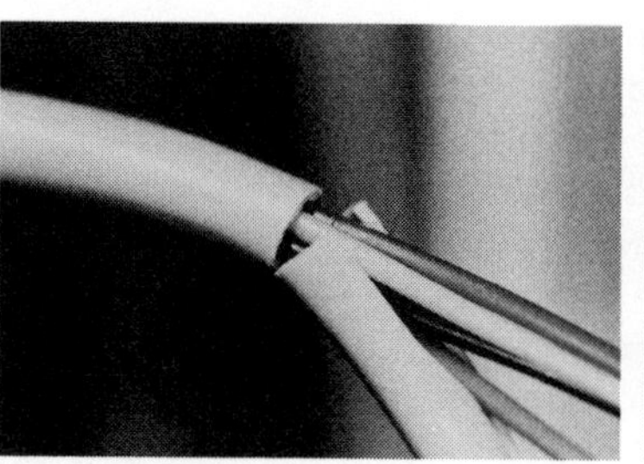

Cut the insulation lengthwise so you can bend the jacket away from the wires inside.

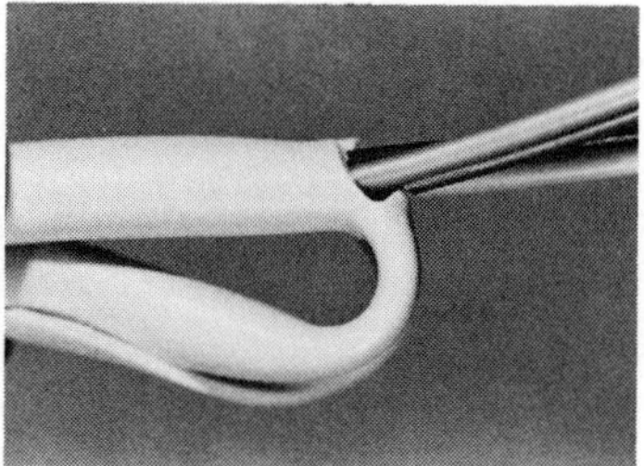

With the outer jacket well clear of the wires, it can be trimmed off with a sharp knife or wire cutters.

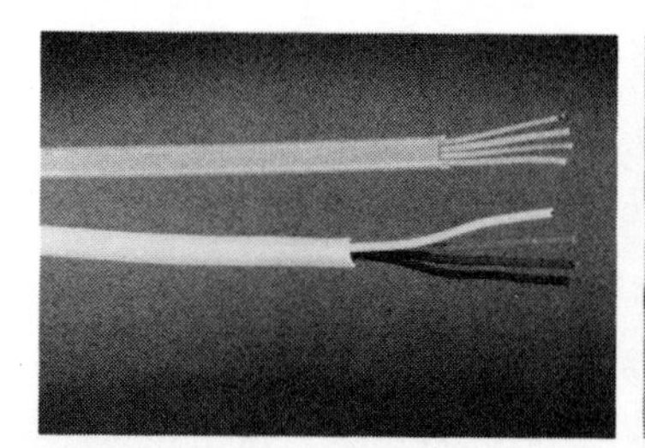

Inspect the four color-coded wires carefully to be sure they were not damaged when you removed the jacket.

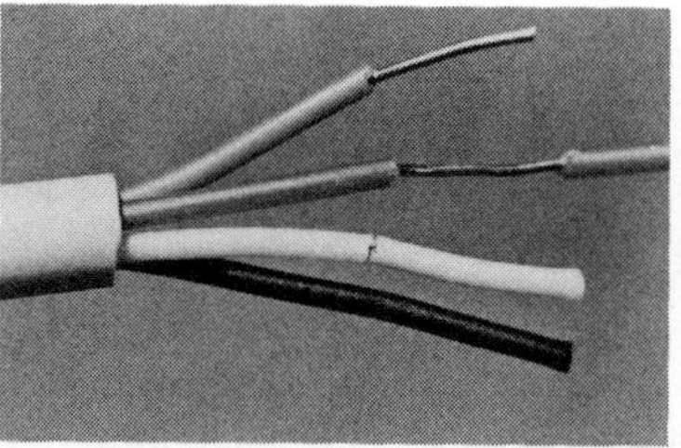

After the insulation on an individual wire is cut with a sharp knife or razor blade, the insulation can be pulled off with your fingers.

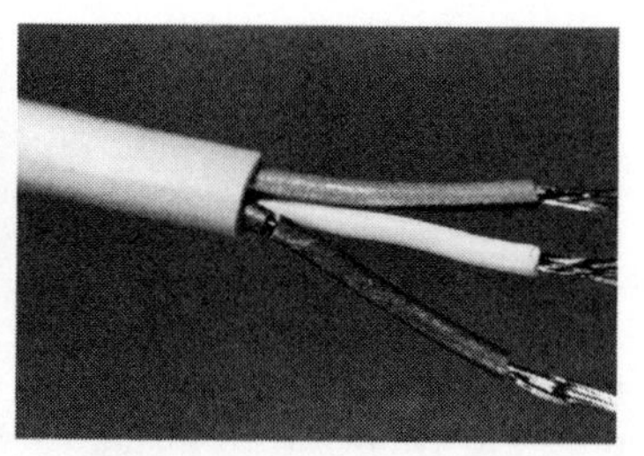

Be careful to cut only the insulation on stranded wire or individual strands will also be removed.

way. Doing one at a time usually results in a tighter fit.

After you have tightened the screw securely, check immediately to be sure the bare ends of the copper wire are not touching other terminals or other metal parts in the housing.

Attaching to Knife Terminals. If the telephone hardware you bought has knife-type terminals, you don't have to strip or wrap the wire at all. These terminals have a V-shaped slot between two sharp metal pieces with the bottom just slightly smaller than the AWG diameter of normal telephone wire. As the wire is pressed into the slot, the sharp edges cut through the insulation and make contact with the wire.

To install a wire on this kind of terminal, leave a long enough piece to grip with your fingertips comfortably. Press the wire into the slot as far down as it will go. You may have to use a needle-nosed pliers or a U-shaped tool to seat the wire completely in the slot. Then use your diagonal wire cutters to trim the end off the wire just past the two metal pieces forming the slot.

This type of connector is widely used in commercial and business installations of telephones where there are multiple connectors for the different telephone lines and services.

These connections are considered permanent. To remove a wire and then reattach it means stripping down and cutting a new length of fresh wire. This type of connector is not designed to accommodate the full range of wire sizes that a screw terminal will handle. It will, however, be the correct size for most of the wiring you'll buy at hardware or telephone stores. These terminals are designed to accommodate only one wire per slot. In almost all cases, there are at least two slots internally hooked together so you can attach your incoming wire to one and your outbound wire to the other.

Attaching with Snap Connectors. Some modular hardware will terminate the ends of the wires in a snap connector, much the same as you use on a 9-volt battery. These are designed to mount directly over the original telephone company installations without having to loosen the wires or disturb the connections. You simply match the color-coded wires and press the snaps on the terminal screws.

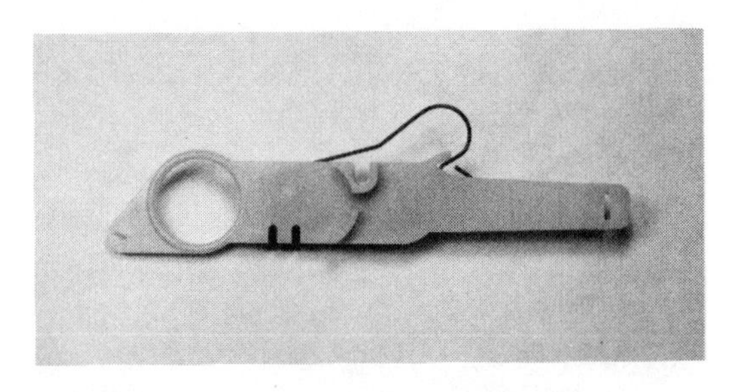

This inexpensive tool is sold for telephone wire installation. It has notches to strip the common telephone wire sizes, the outside insulating jacket of the wire, and measuring scales to strip and leave the correct amount of bare wire for a proper connection.

Working with Modular Extensions

If your wiring plan requires only that you use prewired extensions from existing jacks to complete your home telephone system you may not have to be that concerned about color-coding matches. Hopefully whoever installed the jacks has faithfully followed the color-coding scheme!

Attaching Phone Cords. As a rule, the cord supplied with the telephone instrument (be it a TAC or some more expensive kind) will terminate in a modular male plug. Since modular wall outlets are always female connectors, this means you can plug the phone in without any further ado.

Wrap the bared end of the wire clockwise around the screw terminal. When the screw is tightened, any movement of the wire will force it to fit more snugly against the screw for a good electrical connection.

There is virtually no way to make a wiring mistake when you use a modularized RJ-11 cord to hook a phone into a wall outlet. But suppose the cord you have isn't long enough for the phone to reach where you want it to. No problem. You don't necessarily have to replace the short cord with a longer one. It may be easier and cheaper to link the two shorter cords into a longer one. You use a modular adapter to connect the two cords.

If the two cords end in modular plugs (most do), you buy a small modular adapter, which puts two female connectors back-to-back. Plugging the two short extension cords into the adapter links them into becoming a longer cord, and you can be sure the correct telephone wires are connected inside.

Using Modular Plugs. Once the modular plug is correctly inserted into the connector, a small plastic latch literally hooks the two together. Pulling on them won't separate them — unless strong force is used, in which case, the two pieces might actually break before the latch separates.

To disconnect the modular plug from the modular connector, you have to depress the latch before pulling the units apart. That unhooks the two without damage. Occasionally you have to resort to a thin screwdriver to help depress the latch. This happens when the plug is recessed into a phone case.

If you wrap the wire counterclockwise around the screw, any pressure from the head while it's being tightened will move the wire away from the screw, resulting in an electrically weak connection.

When attaching two wires to one screw terminal, wrapping and tightening one wire at a time gives a more secure fit and better electrical connection.

One of the more useful adapters is the Y-adapter. This permits a single wall outlet to feed two different phones at once. You insert the male plug on the adapter into the female wall outlet and then connect your cords by their male plugs.

Cautions with Modular Installations. Be careful of two things as you go about installing modular extensions. Remember the caution about the number of phones you can operate on a single line in a home: The REN of all the phones in your home should not add up to more than 5.

And be careful that the modular extension cords don't become a safety hazard for walkers, crawlers or vacuum cleaners. Try to keep from using cords longer than you really need. Otherwise, you'll have a lot of extra wires coiled up just waiting for somebody's foot. Keep in mind too that long cord runs will reduce the number of phones you can have on a single line. The low levels of electric current on telephone wires can only go so far so often!

Protecting Your Wires and Phones

Unpredictable — and invisible — power surges or voltage spikes in the electric current of an ordinary AC wall outlet can destroy internal computer components. Most computerists quickly learn to feed electric power to their computers through a surge protector. These devices filter the regular current to protect the computer from harmful power aberrations.

Many times it's easier to work with a longer bared end of the wire to wrap around the screw and then cut off the excess after the screw is tightened.

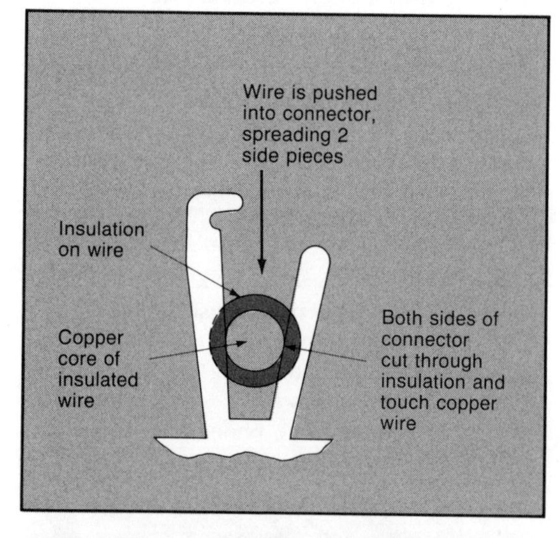

With this kind of terminal, the knifelike sides cut into only the insulation of the wire, making secure electrical contact with the copper conductor inside.

After the wire is pressed snugly into a knife connector, the end is trimmed off.

Knife-type connectors are found in large terminal banks in commercial telephone installations.

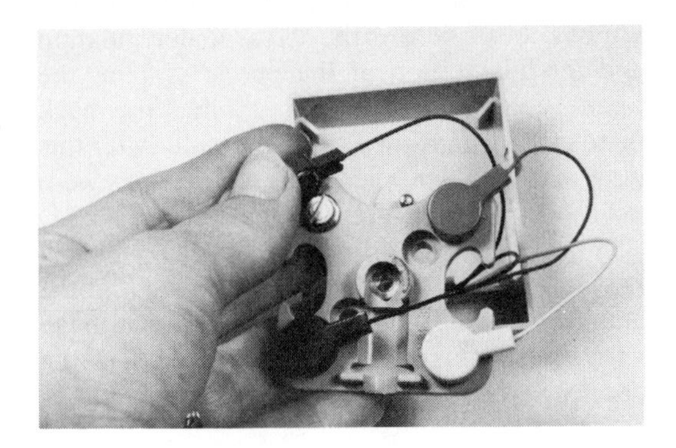

These snaps press on the head of the terminal screw so you don't have to loosen any connections to hook up the adapter.

What most of these same people don't realize is that this kind of damaging voltage spikes can also reach their computers — and other telephone equipment — through phone lines!

While the need for protection in the telephone lines for computers, with their expensive and extensive electronic circuitry, may be obvious, what many people fail to realize is that most telephones — even TACs designed to be thrown away when they go bad — have some of the same integrated circuits and microchips that many computer circuits do.

To protect all of your phone equipment, you would be well advised to install a surge protector specifically designed for phone lines. Install one of these close to the head end of your system. One version of a telephone line surge protector plugs into a normal grounded AC outlet (the two prongs are insulated, only the larger prong is metal for contact with the ground circuit in the outlet). On the side of this device are two modular jacks. The incoming line from the phone company plugs into one; the line to the rest of your system into the other. Inside there are the circuits to suppress large surges of current or voltage spikes.

The small plastic latch on the top of a modular plug hooks into the jack to ensure a tight, electrically secure connection.

A telephone line surge protector like this will protect your phones and accessory equipment from sudden power surges that can happen on phone lines, especially during electrical storms.

After you've connected all of the hardware, plug in your phones and enjoy the convenience and luxury of a custom-designed, do-it-yourself phone system.

13 Hang Up the Phone

Connecting a wall phone to your system is no different, electronically, than connecting any other phone or outlet. The topic deserves separate attention here because of the carpentry techniques needed to fasten the unit to the wall securely. Locating a wall phone also requires more careful attention to safety precautions than phones in most other places.

Types of Wall Phones

There are two basic kinds of phones you can hang on a wall. A true wall phone is designed for a wall mount and nothing else. There are also, however, desk phones that have wall brackets or optional mounting devices that allow them to be hung on a wall.

True Wall Phones. Newer models of the true wall phone will have a "slip-hole" cut in the back to slip over the large head of a mounting bolt on a wall-mount plate. In older model wall phones, built before the days of deregulation and modularization of the phone system, the phone company screwed or bolted the back plate of the phone directly to the wall. After this plate was installed, the wiring connections were made and the phone reassembled on the wall.

Some of the newer modular wall phones will have a very short modular cord to plug into the jack of the wall mount plate. Others will have wiring terminals (screw or knife type) in the phone to which you attach your four-wire color-coded cable directly.

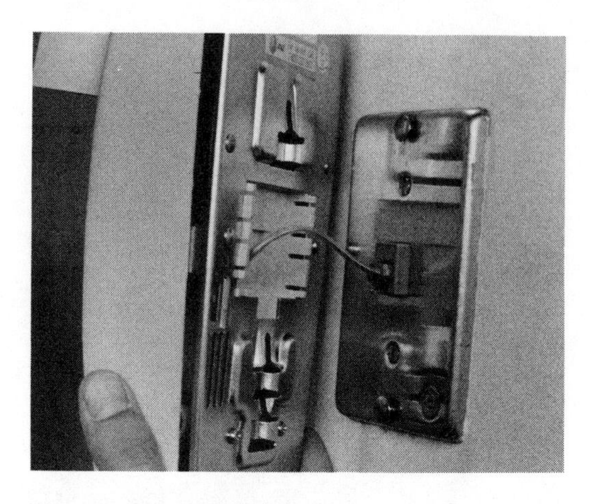

Newer modular wall phones slip over special shouldered bolts on a mounting plate.

Before deregulation, the phone company installed your wall phone by screwing the unit directly to the wall.

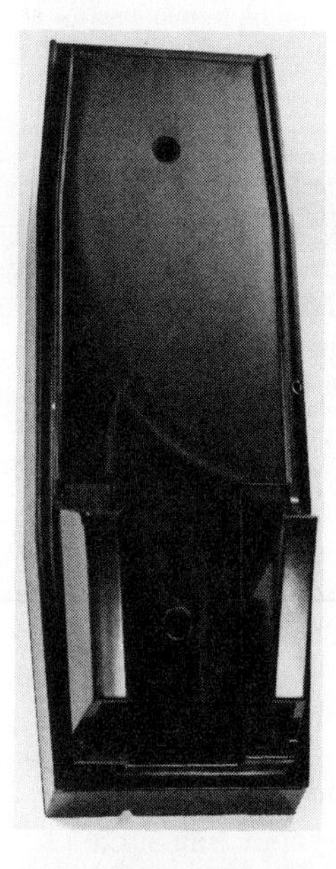

Some inexpensive one-piece phones have brackets like this one, which you screw to the wall to hold the phone.

Modified Desk Phones. Some of the one-piece TAC phones (those "throwaway cheapies") will come with (or offer as an optional accessory) a bracket that you screw to the wall on which you hang the phone instead of placing it on a desk or table. These phones connect to the modular jacks with a cord the same as any other desk phone.

Some of the special-feature phones are built to be either desk phones or wall mount phones. As wall phones, these are connected to phone lines in the same way as you connect them if they were used as desk phones — with the standard modular cord.

Locating a Wall Phone

The location of wall phones must be planned carefully to ensure the most solid support for the phone and the most convenient place for its use and to avoid locations that place you or your phone in any danger of being damaged.

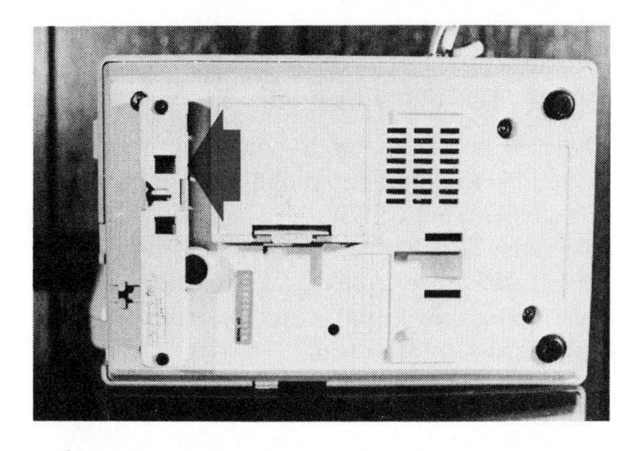

Many of the phones you buy today have an adapter built into the base that lets you mount the phone on the wall or use it as a desk phone.

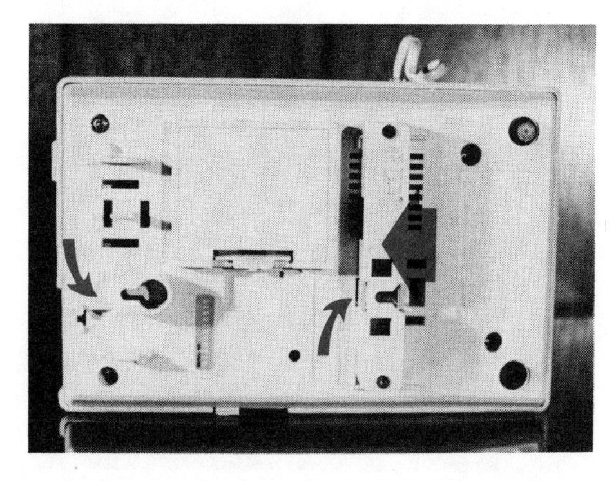

In this photo, the adapter is placed on the phone to let you mount it as a wall phone. The slotted holes attach to the wall phone plate.

When you planned your phone system, you should have located the phone on the most convenient wall in the room. You have to fine-tune that decision when you begin the actual installation.

Decorating Considerations. Installing a wall phone plate usually involves making more holes in your walls than you make when installing only a phone jack. These holes are at a place where any patched mistakes will be more readily seen.

The installation is usually more permanent than other types of phone installation only because it is harder to patch the holes and repaint the area of the wall if the phone is ever moved. Double-check your wall location carefully, and plan all the installation steps carefully so you don't have to do more carpentry and wallboard work than you planned or need to.

Height. How high up the wall the phone should be installed depends on how you intend to use it when you place or get calls. If you will be standing when you dial or answer the phone, a good height is to center the dial about 54 in. from the floor. You may want to adjust this a few inches depending on the average height of people in your family.

Remember, too, a phone 4½ feet from the floor is beyond the reach of some of the younger members of your family. This can be a great advantage if your intent is to keep the phone out of reach of curious minds who may "let their fingers do the walking" farther than your long-distance budget provides for. It can also be a great disadvantage for them, however, if they need to use the phone to reach fire, police or a neighbor in an emergency.

If you plan to use the wall phone while seated on a stool or in a chair at a desk, you would, of course, want it lower on the wall.

A good rule of thumb to follow in deciding on the height is that your arm should be bent at the elbow at a comfortable angle both in using the dial pad and in reaching for the handset. You shouldn't have to reach or stoop to use the phone.

Fastening Considerations. Once you settle on a convenient height, then you have to adjust the horizontal location at that height for the type of installation you'll make. If you've planned to use a receptacle outlet box recessed in the wall as your mounting surface, you'll have to find a stud in the wall and locate your phone on either side of it.

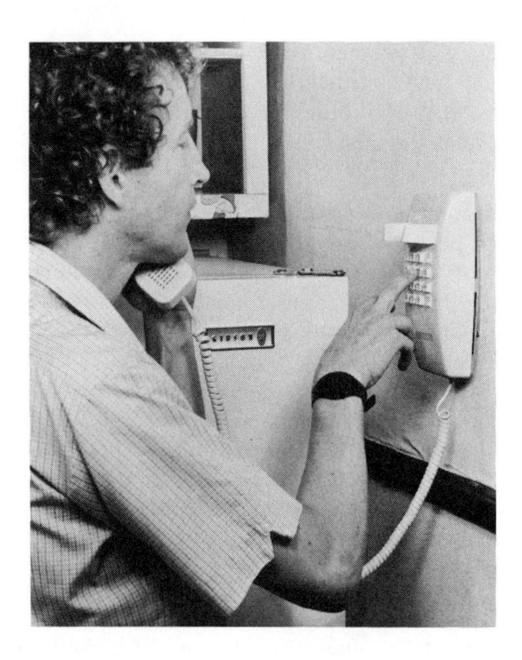

The correct height for a wall phone lets you dial with your arm bent in a comfortable angle at the elbow.

If, on the other hand, you intended to use other fasteners to put the phone on the wall, you'll have to locate the stud to *avoid* it. You don't want suddenly to have to drill or chisel away a large part of a wall stud to make room for the wiring connections that project from the back of most wall phone mounting plates.

Locations to Avoid. Wall phones pose no greater threat to man or machine than any other kind of phone — but their cords do.

If you're installing the phone in your kitchen, avoid any wall location where the handset cord will be stretched over the burners of the stove to reach the place from which some family members seem to prefer to talk. No matter how carefully you plan the location of your phone, some time someone will want to avoid the ambient noise level of a kitchen at meal preparation time, or your teenager will want to duck around a corner for privacy in talking to that "special" calling party.

If your wall phone is destined for your workshop or hobby area, you'll want to be sure the cord doesn't cut across the path of any power tools. Even turned off, the teeth on a radial-arm saw blade can do serious damage to a phone cord pulled across it. If you plan to locate the phone near a tool bench, be sure the cord won't get buried under the inevitable clutter of tools or workpieces on the bench.

Wall phones in the kitchen should not be located where the cord will stretch across the burners on your range.

Some people have found that a wall phone in a garage is the ideal location for their work and leisure life-style. Remember, though, that if it's convenient for you to use in the garage, it's just as convenient for people you don't know if you leave your garage open and unattended.

Fastening Phones to Walls

There is no great weight that has to be supported by a wall phone mounting plate — most pictures on your wall weigh as much if not more than your wall phone. Secure mounting is essential, however, to take the stress of use without ripping the phone — and a large part of the wallboard — off the wall.

Keep all phones in the workshop away from power tools and don't bury the cord under workbench debris.

Wall phone mounting plates have screw holes that match the mounting lugs of a standard receptacle box. The other bolts on the plate are used to hang the phone.

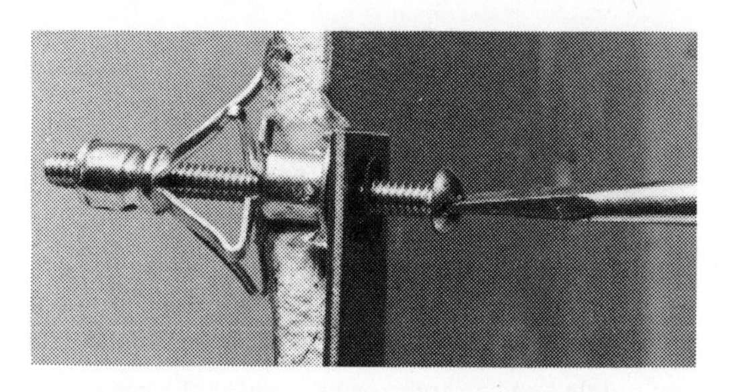

Expansion bolts (or Mollys) have a slotted shield that expands behind the wall as the bolt is tightened, clamping it in place. The bolt can then be removed without the anchor falling out. You can put the Mollys in first and then remove the bolts to install the phone plate.

Receptacle Box Mounting. Most of the modular wall phone plates will fasten to the mounting holes in a standard AC electrical outlet box. Since the box is nailed directly to the studs inside the wall (or fastened securely to the wallboard with its own clips), this installation tends to be the most secure and stable. Wall phone plates usually have two sets of holes for mounting. One set matches the lugs on the receptacle box; the other is for the large-headed bolts on which the phone is hung. These bolts will almost always have shoulders to keep them spaced the appropriate distance from the plate when they're tightened down.

Molly-Bolt Mounting. If you're going to put the mounting plate directly on the wallboard, the best fastener to use is the Molly bolt. After you locate where the mounting hole is, drill a hole in the wall. The hole should be the same diameter as the Molly. Then insert the Molly in the hole and tighten the screw or bolt in the center. This spreads the expansion strips behind the wall. Then remove the bolt and you can slip it through the mounting hole on the wall bracket and tighten the unit to the wall. Be sure to drill for *both* Molly bolts before you fasten the wires, or you'll stand the chance of nicking the insulation off the wires when you drill the second hole. Make the hole for the terminal screws on the mounting plate after you install the Molly bolts.

With the Molly bolts, you can easily remove the mounting plate if you ever need to troubleshoot or repair any of the wiring. This is not as easily done with other fasteners.

Toggle-Bolt Mounting. Toggle bolts have a special kind of nut threaded on them. It is a spring-loaded, wing-shaped device. The wings collapse against the threads of the bolt and then spring out to grip the back of the wall when the bolt is turned down tightly. To use a toggle to hang your phone, drill a hole in the wall big enough for the folding wings. Then remove the wings from the bolt and insert the bolt through the wall phone plate's mounting hole. Then thread the wings back on the bolt (facing the right way). Do this for both the top and bottom mounting bolts on the wall phone plate. Be sure you've cut the larger hole to accept the terminal screws on the back of the mounting plate before you begin to install it with the toggles. Fold the wings of both toggles back and insert them into the holes in the wall. The wings will snap out when they've cleared the back of the wallboard. Then tighten the screws and the wings will move up on the bolt to grip the back of the wall. You may have to pull on the bolt to "set" the teeth of the wings into the plasterboard to keep them from turning with the bolt as you tighten it.

Consider toggle bolts a permanent installation. When you loosen them to remove the bolt, the wings fall away into the inner depths of your wall.

Sleeve-Anchor Mounting. You can also use plastic wall anchors with sheet-metal or plasterboard screws, but this installation will not take the same kind of stress as a Molly or toggle bolt. Anchors are designed to provide a grip for screws in walls where screws alone would not hold securely. The sleeve (usually plastic) is ribbed for a firm grip and expands as the screw is turned into it, wedging itself against the sides of the hole drilled in the wall.

In wallboard installation, tap the sleeve into

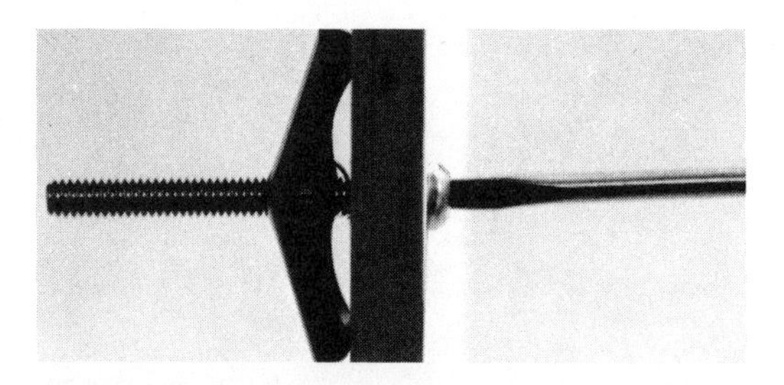

The toggle bolt has spring-loaded legs that compress for insertion in the wall, and then spring open behind it to lock the bolt in place. Because the bolt cannot be removed without losing the legs, the mounting plate must be on before the toggle is inserted in the wall.

a slightly undersized hole and then insert the proper size screw and tighten it.

Sleeves are also made for masonry, and may be the most secure mounting if your phone is going to be hung on this kind of wall. A similar sleeve-type mounting for masonry is a lead anchor, called an expansion shield, which takes heavy lagscrews. These may be too big to fit in the mounting holes of your wall phone plate.

Running the Wires

Once your phone is securely mounted to a safe location on your wall, all that's left is to connect it to your system.

Getting the Wires to the Phone. If you're running wires up the wall from a baseboard jack, you have two choices. You can simply staple the wire to the wall straight up from the baseboard. This is, of course, the easiest way. It does present some problems, however. The wire is more susceptible to damage and it will always be in the way when you need to repaint the room. It's not a good idea to paint phone wires. Some paints can attack the insulation and create phone problems neither you nor the phone company will appreciate. If you look around long enough, you can find different color modular cords or telephone cable insulation that may better match your decorating scheme.

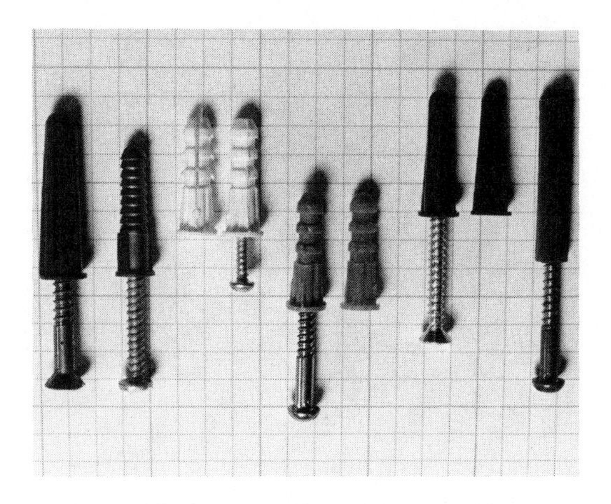

Sleeve-type plastic wall anchors are used with sheet-metal or plasterboard screws and work well in both hollow and solid walls. The sleeve is ribbed for a firm grip and expands as the screw is turned into it, wedging itself into the hole.

A cord tacked to the wall will inevitably be considered very tacky by someone in your family or circle of friends. If this concerns you, a short in-the-wall run is your best answer.

You can drill a hole at the baseboard directly in line with the hole you'll have to drill at the phone location. Then using the "fish wire" technique (described in Chapter 11) run the wire inside the wall for the short length to the phone location.

If your total wiring scheme has already located the wires inside the walls at the correct height, all that's left is to hook up the phone to them.

Getting the Wires on the Phone. Once you're ready to hook the wires to the connections on the mounting plate, the key thing to remember is to follow the color-coding scheme: Always connect red to red, green to green and yellow to yellow. (Other mounting tips you should follow are in Chapter 12.)

14 At the Tone, Leave Your Message

In Chapter 9, you learned about the features you can buy in a modern phone-answering (and recording) machine. In this chapter, you'll look at ways of using these convenient appliances.

An answering machine like this without a phone has a built-in microphone to record your message.

Recording Your Answer Message

First, let's consider the messages you can record for your callers to hear when they dial your number and you're away (or tied up).

Don't Extend an Invitation to Thieves. The first rule here is: don't begin by telling your caller that there's nobody at home. That's an open invitation to a would-be burglar. Instead, say something to the effect that you are "unable to answer the phone in person right now." Then the caller isn't absolutely certain that you're away from home. You might be taking a bath, mowing the lawn or doing your income taxes.

Some people like to begin their answering messages by giving their own phone number. This lets the caller know that they have — or have not — reached the line they wanted to. (Answering machines get as many wrong numbers as plain phones!)

This is a better idea than giving your name. A random caller — with malice in mind — can dial your number without knowing who you are and where you live. If you volunteer your name, it isn't much trouble for the troublemaker to look your address up in the phone directory. Then your home may become vulnerable to who knows what mischief.

This answering machine has a phone you can use to record your answer greetings.

Give Clear Instructions to Tell the Caller What to Do. If your answering machine is one of the many models that has a recording circuit built in, you will want to urge your caller to leave a message for you on tape. So you will want to include a brief instruction that explains how the caller can leave a recorded message on your machine.

All models use an audible beep on the line to signal the caller that the recording circuit has become operational. Therefore, you should always alert your caller not to begin speaking until after the beep has been heard. Otherwise, you'll end up with only the tail end of the message on tape.

Time for the Caller's Message. If your machine has a VOX feature, you needn't really worry about cautioning your caller to be brief. A VOX-equipped model will continue recording the caller's message until there is a disconnected line or until there has been a very long silence on the line. It's a worthwhile technical feature of the better models.

On the other hand, if your machine has a fixed recording interval, your answering message should alert your caller to it. Otherwise, the machine may cut off before the caller has finished speaking. This is why we recommend a machine with VOX. A short, fixed interval is apt to make the caller jittery about making the message fit the time allowed.

Information from the Caller. You will find that it is highly desirable for you to have the name of the caller as well as the call-back number. If your answering machine is programmed to operate over a stretch of several hours or days, you will also discover that it is useful to know at what time the call came in. (A few fancy answering machines add that information automatically to your cassette with an audible "time stamp.")

It is unwise to ask the caller to mention the day or date of the call. That's a dead giveaway that you expect to be away from home for an extended period. It would be almost as bad as allowing newspapers and milk cartons to pile up on the front stoop during your vacation.

Sample Message Greetings

Here then is a sample of how your answering message might be composed:

Hello. This is 555-1234. We're busy right now and can't answer in person. After you hear a tone on the line, please leave a message (of no more than 30 seconds) for us. Include your name, phone number and the time of your call. We'll get back to you as soon as we can. Thanks.

Messages for Answer-Only Units

If your machine is a simple "answer-only" model (that is, one without recording capacities), your message will merely advise your caller to try again later. Such a message might "read" like the following example.

Hello. This is 555-1234. We're busy right now and cannot take your call in person. Please try us again after 3:30 this afternoon. Thanks.

Keep Your Answer Greetings Current

It is probably obvious that an effectively composed answering message used on a machine with recording capability can be used over and over for years. That is not true with messages prepared to play back through an answer-only model. In the latter case, you will have to modify the callback time to suit each occasion. Otherwise, the machine is really not much more helpful to your callers than an unanswered dial.

If you own an answer-only model, better make it a rule not to operate it when you have to be away more than eight to ten hours at a time. If your message indicates that the caller should phone again, say, "tomorrow" or "next Monday," you are exposing your home or apartment to a risk — should a mischiefmaker call at random or on a hunch.

Prerecorded Answer Greetings

Some people think it's better to have their machines answer with a professional touch. They want an impersonal voice or a bit of humor to greet their callers.

Many stores that sell answering machines also sell prerecorded answer greetings. These vary in style from a professional radio announcer reading a straightforward message all the way to musical jingles and comic impersonations of famous people. Some tapes contain a set of different messages through which you can rotate for the sake of variety. There are even sets with seasonal messages for holidays. (Of course, you can also add holiday greetings to your own homemade tapes if you wish.)

These "canned" messages can't include your own telephone number, but they can — and should — contain the other essential elements mentioned previously. Be sure to listen to the message(s) before you buy them. They should be clear and concise, however cute or jazzy. It's all a matter of personal preference — and budget. The prerecorded messages do add cost you don't have when you do your own recording.

Strive for Good Sound Quality

One final point about recording answering machine messages of your own. Be sure you do your recording where you will have good sound quality. Don't make the recording in an environment with a lot of background noise.

Good Quality Begins with Good Mikes. If your machine doesn't have a built-in microphone for this purpose — or if you don't have a standard cassette recorder with its own mike — then record your message through the best phone you own. An instrument built into the machine is likely to be of acceptable quality. Don't use a TAC telephone as the microphone. The voice quality is apt to be substandard.

Speak Clearly. As you make the recording, speak clearly but naturally, keeping your voice

Keep your mouth about 12–24 in. away from a built-in mike and speak in a normal voice when recording an answer greeting on a unit like this.

volume neither too high nor too low. Be careful not to place your mouth too close to — or too far from — the mike. If it is, the recording will sound distorted or distant.

Preview Your Message. Listen to the taped message played back through the phone before putting it into regular service. Be prepared to redo it if it isn't your high quality.

Quality of Cassettes Is Important

One important point to keep in mind is that tape cassettes — in regular sizes or the tiny microcassettes — are available in a wide range of lengths and qualities. Cheap cassettes are likely to give your machine problems. They may sound distorted and noisy. What is far worse, they may tangle up in the mechanism, destroying themselves and temporarily disabling the machine itself. Buy only good voice quality cassettes from reputable manufacturers. Avoid off-brand bargains.

This doesn't mean you must purchase "audiophile quality" cassettes used in recording high fidelity music. Such expensive tape types would work, but you would be paying for quality your machine doesn't require. In general, look for tapes that are suitable for "normal" (120 microsecond) recording. Don't choose metal or chrome tapes designed for "high" 70 microsecond recordings.

Match the Cassettes to Your Machine

When choosing cassettes for recording, be certain you pay close attention to the advice given in the user's manual for your particular model.

Microcassettes. If your model requires the tiny microcassettes, be sure you choose a unit labeled in exactly that way. A similar tape form called a "minicassette" is often sold in the same stores. The two forms are not interchangeable. (Minicassettes are used almost exclusively for inexpensive dictating machines. Their sound quality is too poor for phone applications.)

You might like to know that a microcassette, just like a larger regular cassette, is designed for recorders using capstan mechanisms. Minicassettes depend on the old-fashioned rim-drive machine design.

Leaderless Cassettes. Some machines require leaderless cassettes or endless-loop varieties. These special types may have to be obtained from the store that sold you the machine. A leaderless cassette is one that does not have the clear plastic section at the start and end of each side of the tape itself. Because there is no magnetic recording compound on this clear leader section, no sound can be recorded there. This means that you must advance the tape past the clear plastic leader before placing the cassette in service inside your answering machine. That's also why many standard cassettes designed for use in answering machines are leaderless cassettes.

Endless Loops. An endless cassette is arranged so that it simply goes around and around in a loop of some fixed duration, say 10 minutes. Experts tell us that these more expensive forms suffer more from wear and tear than regular cassettes.

Be sure you buy the right size cassette tape for your answering machine.

Tangled audio cassette tapes can be repaired with a little patience and a minimum of tools.

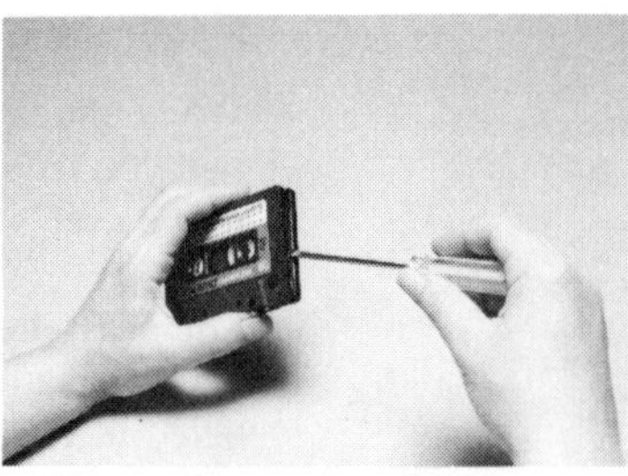

Lift one side of the cassette with a small screwdriver and gently open the cassette chamber.

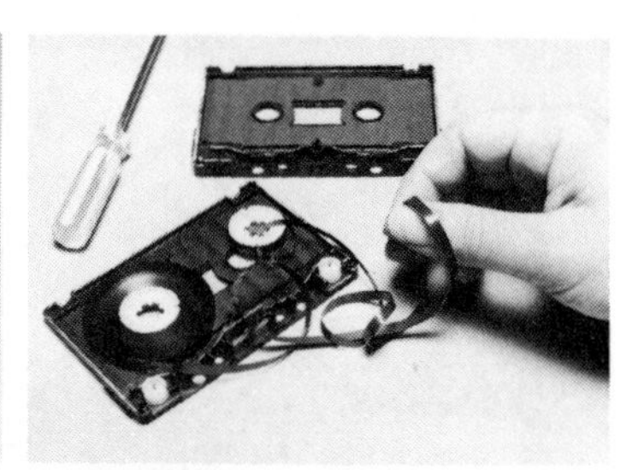

Pull the tangled tape out of the cassette without stretching or breaking it.

Use a pencil in the cassette hub to wind the tape carefully on one of the internal reels.

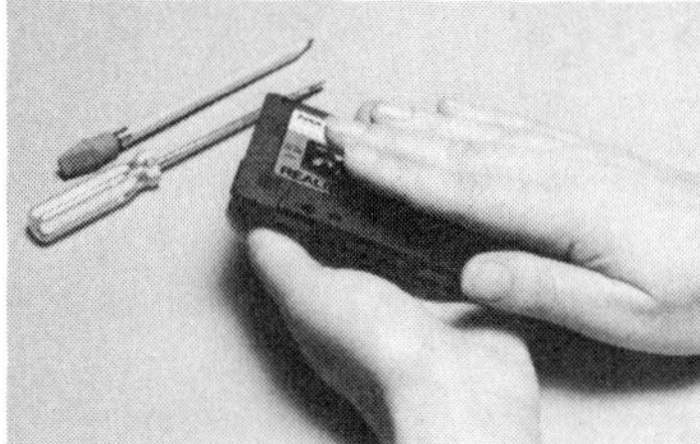

Give the reassembled cassette a good slap with the palm of your hands to free the tape.

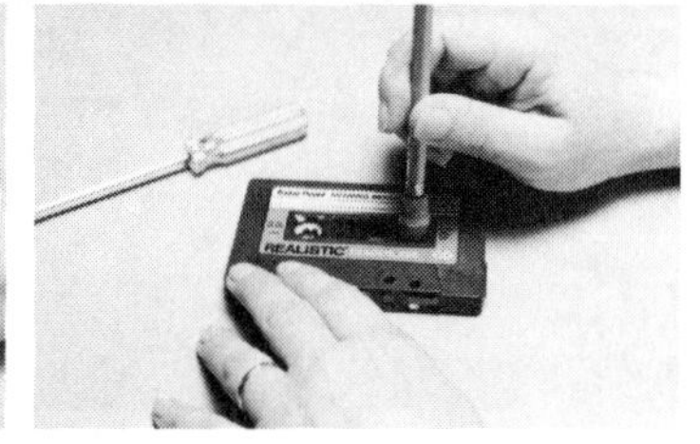

Before rewinding cassette in machine, use a pencil again to wind tape tight against each hub.

Listening to Your Messages

Let's now consider the best ways of listening to the recorded messages from persons who called while you had the answering machine on.

The process nearly always requires that you have paper and pencil handy to jot down the names, numbers and other information you hear on the cassette message tapes. Don't make the mistake of trying to keep track of the calls in your mind. It's almost inevitable that you'll forget one, especially if you have quite a line of messages waiting for you in the machine.

Try to locate the answering machine so that the playback controls are handy. The top of a desk is ideal. Making notes while reaching back to a machine "hidden" away behind a row of plants or beyond a barrier of shelved books will lead to all sorts of frustrations and problems.

This is because you will find that you often have to instruct the machine to repeat individual messages, sometimes several times. Many callers speak too rapidly or indistinctly for you to catch the vital information on first hearing. This requires a convenient hands-on relationship to the machine controls.

To avoid confusion, you should make it a strict practice to erase old taped messages before putting the recording cassette back into service. Otherwise, you run the risk of running into old messages (left on the tape) without realizing it. While some rather expensive models will take care of the erasing more or less automatically, most depend on the user's judgment.

Unless you own a model that deletes false starts on the tape, be prepared for one sure source of irritation most people experience with their answering machines. If a caller doesn't choose to leave a recorded message, the machine will leave a trail of noisy stops and starts for you to skip. That's another reason for putting the playback controls at your fingertips.

You may well find that about 50 percent (or more) of your callers hang up before the beep even comes onto the phone line, Even so, most models have no way of detecting this fact until the recording circuit is activated. The consequences are annoying. You can see why some people prefer to have answer-only models.

Call Screening

You may decide to use your answering machine as a call screener when you're at home. You let the machine go through its motions, while you listen silently on the machine's extension. If the caller is somebody you really want to talk to at that moment, you interrupt the recording and turn the machine off.

There are models on that market that permit

Most standard audio cassette tapes have blank leaders spliced to the front that will not record any sound.

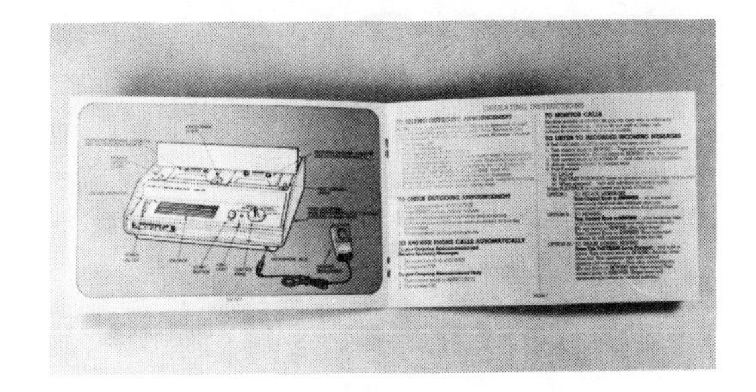

Don't ignore the user's manual, which has many tips and suggestions on how to use your answering machine.

a caller to signal the machine to alert you that he or she is on the line. Obviously, you must supply those privileged callers with the signals to use. This feature is especially nice for other family members to use.

Electronic Memos

Speaking of family members brings to mind the fact that an answering machine can be used as an electronic memo box. Say, junior wants to let you know that he's staying late at band practice. He calls home and, after the recording beep cues him that the recording circuit is operational, he leaves that news on the tape for you to hear when you arrive home from work. No callback is required.

Close friends and business associates often use answering machines for aural memos too.

As a matter of fact, you may well discover that it's a handy way for you to leave memos to yourself as well. Rather than writing down some essential piece of information on a slip of paper in your crowded pocket or purse, you dial your own number and speak the information into the recorder. Then when you get home, you can act on it appropriately.

If you do a lot of your business calling from your home telephone, you may want to use the answering machine recorder as a sort of revolving file of calling records. Rather than reusing the same message cassette over and over, you would keep several dozen blank cassettes on hand. Then you would number and use them sequentially, saving the messages recorded on a tape for a couple of weeks as a temporary record of calling transactions. Really important messages could be saved for longer periods by being "dubbed" off onto a file tape. (This would require a simple "real-time" cassette duplicator of the sort now widely sold in hi-fi and appliance stores.)

If you elect to use your machine in this way, do keep in mind the point about buying only high-quality cassette tapes.

Using Special Features

Your answering machine may have special features like call forwarding. (Review the list of special features in Chapter 9.) To use them correctly, we suggest you review your user's manual. In fact, that manual should be your steady companion until you know exactly how to operate your answering machine. Don't misplace it. And take our best advice: don't try to use a new machine without first carefully reading through the manual. Not only might you do things wrong; you might even damage the device itself.

15
Bleeps, Bloops and Bauds

Even an experienced computerist faces the challenge of a strange new world of jargon and specifications when installing a modem for the first time. To communicate by telephone lines you need a modem and a data terminal. But your home or business computer is *not* a terminal. All of the complicated steps involved in hooking up a modem involve "fooling" the computer to think and act as if it were a data terminal.

Most people who are experienced with their home computers will read the instruction manuals only when all else fails. Don't be tempted to do this when you install a modem for the first time. Armed with the instruction manuals for your modem, computer and communications software, follow these installation steps:

- Set up the modem
- Connect the modem to the computer
- Configure the communications program
- Check compliance with phone company and FCC regulations

Set Up the Modem

When you unpack your modem, save all of the packing materials and the box it came in. There are no parts inside most modems you can (or should) service yourself. The original packing will be the safest material to use when shipping the modem back to the manufacturer or taking it to an authorized service center if something goes wrong.

Check the Controls and Connectors. There will be some controls and connectors on all modems. Plug-in modem cards for such computers as the Apple or the IBM-PC and its clones will usually have only an RJ-11 jack and maybe a DIP switch to preset some modem functions. Stand-alone modems will also have a DB-25 pin connector that goes to the RS-232C port on your computer and some sort of connection to the AC power. You may also find a volume control for the internal speaker and an on–off switch.

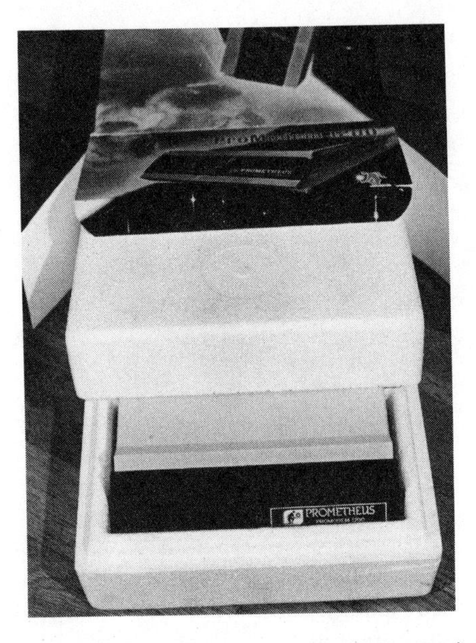

Save the box and packing your modem came in just in case you have to return it to the store or manufacturer.

Properly identify and locate all of these modem controls and connectors (using your instruction manual as a guide) before you go any further with the installation.

Finding a Place for Your Modem. If you have a stand-alone modem, you'll have to clear a space for it and your telephone on your com-

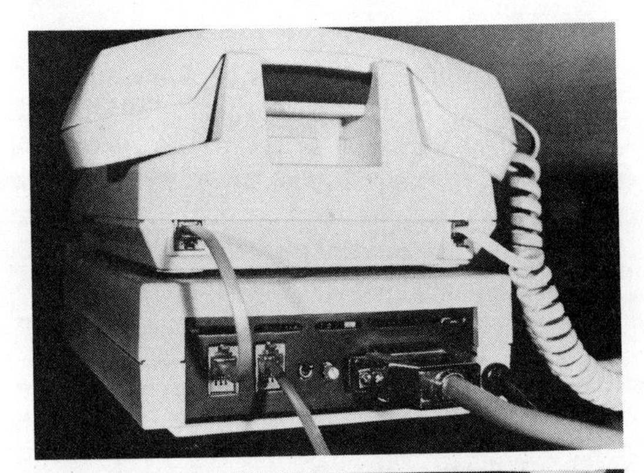

Before you start installation, take the time to locate and identify all the controls and connectors on the modem.

puter work area. Most stand-alone modems are the same size as a desk telephone base and so you can put your phone on top of the modem. Some people use double-face tape or picture-mounting adhesive squares to put the modem on the back, side or top of their computer's display screen console.

If you have a plug-in card modem, you'll have to remove the case from your computer's CPU unit and carefully insert the card in the appropriate expansion slot. Some computers have a slot reserved for a modem card with different connections than other expansion slots. Check your computer's instruction manual carefully, and follow all directions for this kind of installation furnished with your modem. It is critical that the modem card be seated properly in the expansion slot so the multipin connector on the end of the card doesn't make contact with the wrong leads in the computer.

Connecting to the Power. All of the card modems, and many of the plug-in modems for computers like the Commodore 64™ or 128™, draw their AC power from the computer itself. You don't have a separate power cord to worry about.

All stand-alone modems must be connected to a source of AC power. Some (like the Hayes Smartmodem™ and others) use low voltage from a transformer that is very much like the "battery eliminators" you buy for portable radios or small tape recorders. These connect to the modem with an ⅛-in. or power cord adapter plug. Don't substitute other battery eliminator transformers for the one that came with your modem. There are different current and voltage ratings — even on the so-called universal models. You can quickly destroy your modem

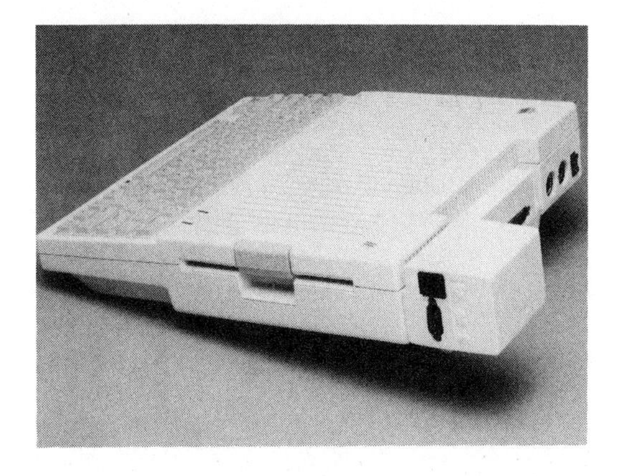

Modems like this plug directly into the computer and draw their power from the computer.

investment with the wrong voltage or polarity. Other stand-alone modems have internal power supplies that simply plug into a properly grounded wall outlet with a three-prong AC plug.

When you get to this point in your installation, seriously consider running any external AC power to your modem (and computer) through a surge protector. These devices protect the delicate internal circuits from damage caused by power line aberrations like sudden voltage increases or high-voltage spikes.

If you bought a modem that acts independently of the computer with a communications buffer to send and receive data unattended, you may want to connect the modem to its own source of AC power, so that you can leave it on while other components of your system are turned off.

Connecting to the Phone Line. Virtually all modems will connect to your phone line with a standard RJ-11 modular jack. If your modem has only one RJ-11 jack, you will want to use a modular Y-adapter in this outlet. One jack of the adapter will go directly to your telephone wall jack with an extension cable, the other to the telephone you'll want to have in your computer work area.

Many modems now have two RJ-11 jacks

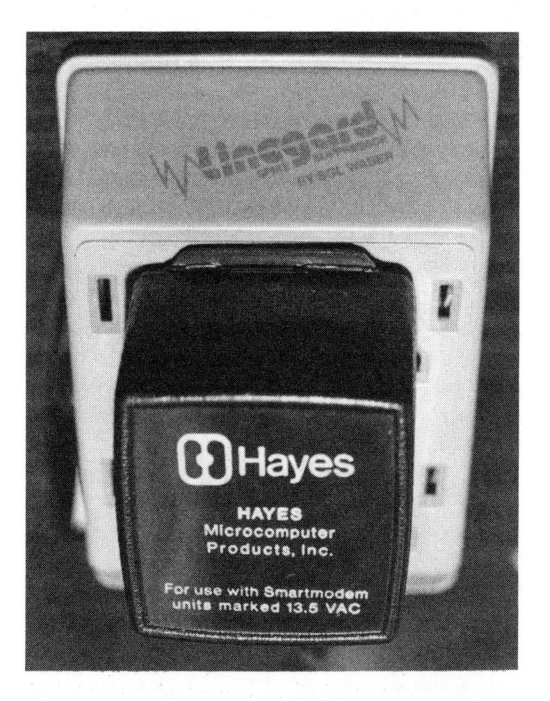

It is a good idea to draw power for your modem through the surge protector you use for other components.

built in, one for the connection to the line, the other for your phone. On many of these modems, the phone will automatically cut out when you begin communicating in the data transfer mode. When the modem's not in use, these two jacks link internally, letting you use your phone as a normal extension.

Configuration Switches. Some modems come with a DIP switch to configure your modem to the kind of cable you're using to connect to the computer and to preset special modem functions. These switches are small — about 10 of them are lined up side by side in the space of a 1-in. integrated-circuit socket.

Each manufacturer will use a different arrangement of switches — there is no universal standard. Descriptions of what these switches do will be in the instructional manual of the modem you buy. Study the switch arrangement and settings for your modem. Know what each switch does if you have to change the setting.

In almost all cases, these DIP switches will be set at the factory in their "default" position. These are the settings that will most likely work with any computer or communications program. If in doubt, leave them set in the default positions. Change the settings only when the modem doesn't work in your initial test or when specifically instructed to do so in the manuals of your computer or software.

Connecting to Your Computer

Your computer communicates through a device known as the RS-232C communications port. On some computers this is completely inside the CPU case, with no external connections to make. On other computers, the communications port terminates at the back of the com-

Two RJ-11 jacks on the back of a modem let you plug in a phone for voice communication. The phone cuts out when data transmission is going on and becomes a normal extension when the modem's turned off.

These tiny switches are used to set certain modem functions and configure it to your computer and cable.

puter in the RS-232C connector. The "RS" of the RS-232C nomenclature means "recommended standard." The recommendations of the Electronics Industries Association are only loosely followed by the makers of computers and modems.

Handshaking. Only three wires of the 25-pin RS-232C connector are used to send and receive data. The rest (of those actually connected) are used to let the two communications ports "talk" to each other. Among other things important to these devices, they tell each other when one bit of data is received (or sent) and the computer is ready for another. This whole process is known as *handshaking*. The configuration of the RS-232C cable assures that the proper handshaking will take place between your modem and your computer, and that your modem will send the proper signal to someone else's modem to let you communicate.

Connecting the RS-232C Cable. Be sure you have bought the correct RS-232C cable. Simply insert the keyed DB-25 connector into the plug or jack on the back of your computer and your modem. These connectors have tapers on the narrow edges so they can be connected only one way. Almost all of these DB-25 plugs and sockets come with screws on each end of the connector. Tighten these into the appropriate hole in the computer or modem to seat the cable securely and to keep it seated securely if the cable should ever get kicked or pulled.

Modifying RS-232C Connections. If your computer's serial port is wired as data terminal equipment (DTE) then in most cases it can be hooked directly to the modem (most of which are wired as data communications equipment — DCE). A "straight-through" RS-232C cable will

The DB-25 plugs used on RS-232C cables have screws like this to connect the plug tightly into the connector.

provide the correct connections between a DTE computer and a DCE modem. The cables will (or should) also correctly connect the data set ready (DSR) wires that tell the computer that the modem is ready and the data terminal ready (DTR), which tells the modem when the computer is ready. (See the RS-232C diagram.)

If your computer and modem won't let you use a straight-through RS-232C cable, you may have to make some modifications in a connector by jumping leads between some of the pins or sockets. If there are any connections indicated in your computer's manual that don't match the recommended RS-232C standard or those expected by your modem in its instruction manual, check your computer documentation (or your dealer) before jumping any leads and sending voltage where you may not want it to go.

Do-It-Yourself RS-232C. You can make your own RS-232C interface cable. The RS-232C standard connections are shown in the diagram.

The best connectors for these homemade cables are those with push-in pins and sockets. You cut a 25-conductor cable to the length that suits you and then solder each end of the wire to a pin or socket.

The 25-conductor wires for RS-232C serial cables come as either round cables or flat ribbon cables. Each wire is color-coded so you can keep the wires straight. With the round cable, however, watch the dual colors carefully. Manufacturers run out of solid colors with which to code after about 8 or 9 wires, and they then use stripes or patterns of two different colors. All flat ribbon cables have a stripe down the wire on one edge. Ribbon cables are easier to keep straight because you can solder them to pins or sockets in a straight-line sequence.

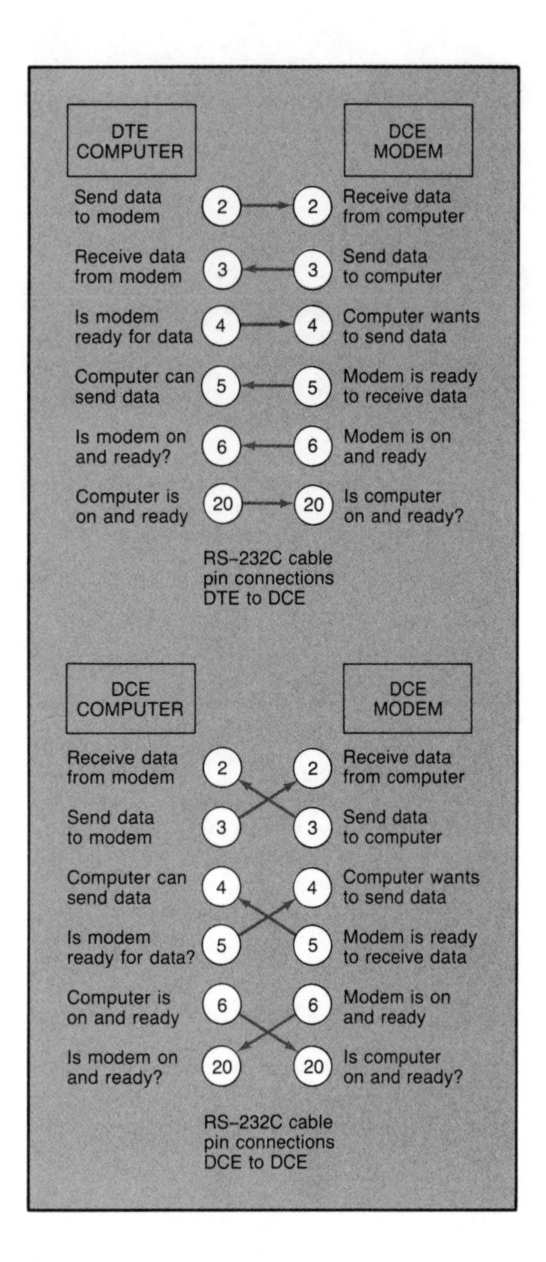

The recommended standard for connections in the RS-232C cable hooks up data terminal equipment (DTE) to data communications equipment (DCE).

Some computers may be wired internally as data communications equipment. Then, this DCE to DCE wiring sequence is used in the RS-232C cable.

Double-check each end of the wire with an ohmmeter after you've soldered a pin on one end and a socket on the other. These pins and sockets then push into an appropriate hole in the connector body. If you need to move wires later, the pin or socket may be pushed out of the body again and you shouldn't have to re-solder anything.

There are DB-25 connectors designed for use

This do-it-yourself connector uses individual pins inserted into the D-type connector. Wires are soldered to the pins before putting them in the connector. Pins can be moved around to change wire configurations.

only on flat ribbon cable that don't require soldering. You press these down on the cable with a special tool or a vise, and the special connector breaks through the insulation of each wire, making contact with an appropriate pin or socket.

Configuring Your Software

Most of the time you will have to configure the communications program to the specific characteristics of your computer. Before you begin to modify any program, or even to use one specifically designed for your computer, make a copy of the original distribution disk. Then carefully store this as a safety back-up and work with your copy both in configuring the program and in using the modem.

RS232-C RECOMMENDED STANDARD CONNECTIONS

Pin 1 Protective Ground. This is connected to the case of the modem or computer which is grounded through the power cable.

Pin 2 Transmit Data (**TXD**). Input to the modem from the computer. This signal is the serial data that is to be transmitted to the modem on the other end of the phone line connection.

Pin 3 Receive Data (**RXD**). Output from the modem to the computer. Serial data received from the modem you're calling is transferred back to your computer on this line.

Pin 4 Request to Send (**RTS**). Input to the modem from the computer telling the modem it wishes to send data. This is not used on all modems and is usually ignored.

Pin 5 Clear to Send (**CTS**). Output from the modem to the computer. CTS is a handshaking line that tells the computer when to stop sending serial data to the modem for transmission. It indicates to the computer when the modem is ready and waiting for more serial data to send out on the phone line.

Pin 6 Data Set Ready (**DSR**). Output to the computer telling it the modem is on.

Pin 7 Signal Ground. Common ground reference potential for the RS-232C signals for both the modem and the computer.

Pin 8 Data Carrier Detect (**DCD**). Output to the computer to indicate when the carrier of the modem you're calling has been detected.

Pin 9 Positive DC Test Voltage (**+V**). Can be used for a modem power test point but normally should not be connected.

Pin 10 Negative DC Test Voltage (**−V**). Can be used for a modem power test point but normally should not be connected.

Pin 11 Equalizer Mode (**QM**).

Pin 12 Usually no connection. A secondary connection for Data Carrier Detect.

Pin 13 Usually no connection. A secondary connection for Clear to Send.

Pin 14 Usually no connection. A secondary connection for Transmit Data.

Pin 15 Transmitter Clock for Timing (**TSE**). Used to send a signal from the remote computer through the modem to your computer to time the transmission of data for synchronous communications. Not used in normal asynchronous communications.

Pin 16 Usually no connection. A secondary connection for Receive Data.

Pin 17 Receiver Clock for Timing (**RSE**). Used to send a signal from your computer through the modem to the remote computer for synchronous communication. Not used in asynchronous communications.

Pin 18 No connection.

Pin 19 Usually no connection. A secondary connection for Request to Send.

Pin 20 Data Terminal Ready (**DTR**). Input to the modem to indicate when the computer is ready for communication.

Pin 21 Signal Quality Detect (**SQ**).

Pin 22 Ring Indicate. Output from the modem to the computer to signal the software program when a ring signal has been detected on the phone line.

Pin 23 Data Rate Selector. Output from the modem to the computer to tell if a connection has been made at 1200 baud or at 300 baud.

Pin 24 External Transmitter Clock (**TC**).

Pin 25 No connection.

On some external plug-in modems, the unused pins in the RS-232C standard are used to feed DC voltage to the circuits in the modem. For this reason, be sure you connect wires to *all* 25 pins.

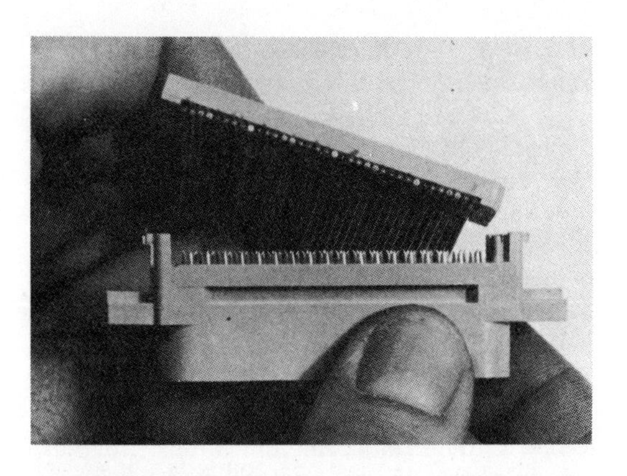

With this connector, you can attach flat ribbon cable to the adhesive surface on the top and then press the units together to make the connections. The pins break through the insulation to make the contact.

Computer-Specific Programs. There are an increasing number of communications programs now specifically configured for one modem with one computer.

There are also several programs on the market now that work only with self-contained modem cards (such as Hayes MicroModem™). Most of these need only superficial configuration if any at all.

If you get a public domain communications program from your computer user group, it will most likely already be installed and all you have to do is unpack it, make a copy and use it.

Third-Party Communications Programs. The widest selection of features and versatility is in the "generic" communications programs made by vendors for any modem or computer. Most of these must have settings made initially before they will work with your computer or modem. Most of these have simple, menu-driven configuration procedures. Public domain programs, not set up for your computer, will require much more extensive configuration — most of the time delving into machine language files to make the changes.

Finding the Communications Port. Your computer isn't really used to sending data to the communications port as it does to your screen, disk drive or printer. It has to be told in your communications program what the address of that port is.

In many computers, a sophisticated integrated circuit called a Universal Asynchronous Receiver/Transmitter (UART) is the heart of the serial communications port circuits. (A common 8-bit UART is the 8251.) The microprocessor receives instructions on what to do with and where to send data by opening up different addresses with coded binary bits. With serial communication through a device like the UART, you can only "open the door" of your serial port by sending that coded sequence of bits from your processor. Other codes will tell the processor to send data to the screen, disk drive or printer. Still other codes tell it to receive data from the keyboard, disk drive or serial port.

Your computer's instruction manual will (or should) identify what this code byte is — usually expressing it in hexadecimal terms such as **EC** and **ED.** The two addresses you should look for in your computer's documentation are the *data port* and the *command/status port*. Some of the weaker documentation will give you this information in binary form although your communications program wants to know it in decimal or hexadecimal form. You'll have to translate the 8-bit binary sequence into decimal or hexadecimal before you begin to install your program. For example, the binary code for data out on a Sanyo MBC-1250 8-bit computer is 11101100. This converts to **EC** in hexadecimal and 236 in decimal $[(1 \times 128) + (1 \times 64) + (1 \times 32) + (0 \times 16) + (1 \times 8) + (1 \times 4) + (0 \times 2) + (0 \times 1) = 236]$.

Asynchronous Communication Parameters

Most software programs will want to know the format of the stream of data bits it will send to your computer so it understands when the transmission of each byte of data begins and ends. This format includes one or more start bits, a data word of seven or eight bits, one or two stop bits and perhaps a parity bit.

Even if your communications program does not ask you for these communication parameters, you should look them up in your manuals and make a note of them someplace. Some of the people you're going to communicate with by modem will have less sophisticated software that requires them to set their program with your parameters. They will need information from you such as "eight data bits, one stop bit and no parity."

Start Bits. The start bit is sent at the beginning of a byte to tell the devices to get ready for the data that follow. If the modem receives the start bit but hasn't finished processing all the bits from the previous set, it will tell the modem on the distant computer to wait before sending any more. Since most of the time only one start bit is sent, you may not be asked for this information to configure your program.

Data Bits. It takes 7 bits of on–off data to make a character in the ASCII coding scheme (see Chapter 5 for an explanation of the ASCII standard). There are, however, 8 bits to a byte of computer data. The eighth bit is used for various purposes on different computers. Some tell whether the computer should display the character normally or in inverse video on the screen. The communications software will need to know if your computer requires an 8-bit or 7-bit data word when it receives and transmits data.

Parity. You will be asked in the communications program if your computer requires odd, even or no parity. The software or firmware in your modem adds up the bits to be sent for a character then puts in the parity bit to make the total come out either even or odd for error checking. Some computers do not require the parity bit to be sent.

Stop Bits. The asynchronous nature of serial communication requires that a signal be sent to tell the receiving modem and computer when all 7 or 8 bits of a data byte have been sent. Serial bits are sent out much more slowly than the parallel bits the computer is used to handling internally. Your computer may require one or two stop bits at the end of each sequence of data transmission. The software provides the stop bits, and so it has to know how many to send out.

Setting Software Defaults. Some of the installation menus of communications programs will let you change the default settings. These default settings are the values the program begins with when you first boot it up. It's best to stay with these default settings until you've used the program with your modem for a while and find you have to change one or more settings every time before you can communicate.

Baud Rate. Most communications programs will boot up at 300 baud. With the faster 1200 or 2400 baud modems you may want to change the default to the faster speed. If one modem does not automatically adjust to the speed of the other modem, changing the boot-up default could save a lot of keyboard commands at the beginning of a communications session.

Duplex. Full-duplex modems can send and receive data at the same time.

A half-duplex transmission goes in only one direction at a time. Each computer can send and receive information in half-duplex, but not at the same time.

Even if your modem can communicate in full-duplex, the modem you're calling may be capable of only half-duplex, influencing your setting through the communications program. You may want to change the default setting of your program if you have to change the duplex with every call.

Tone-Pulse. On modems with the autodial feature, you can select if the dial signals will be sent out to the telephone line as rotary dial pulses or electronic tones. Most communications programs come with electronic tones as the default setting. If your telephone line is still in a pulse dial system, you'll have to change this setting before you can dial out through your modem from your keyboard or automatic dial telephone directory in your program.

Other Settings. Some of the communications programs will let you modify the default settings for various features of your modem. On some, for example, you can change the default setting so your auto answer feature will answer the phone on a specific ring — say the fifth ring. Other features you can change include whether the modem will automatically redial a number that's busy or just quit and let you do it. You should be able to adjust, through software default settings, such things as screen echo of characters when you're not hooked up to another computer, whether you disconnect after a file transfer or switch to the terminal mode, or whether your modem will automatically stamp each session with the time and date and the time you've been connected.

Check Compliance with Phone Company and FCC Regulations

Your modem, like any other equipment you hook up to your telephone system, must have the appropriate FCC certification and REN. You are required to report these to the phone company to let them know what kind of equipment is feeding into their lines. If you don't, the phone company can cut off your service.

16
Finding Phone Problems

Even though most telephones can be repaired "in theory," there is often a question of whether the phone is worth the cost or effort involved.

A broken TAC (throwaway-cheapie phone) is usually not worth taking apart to repair unless the problem is both minor and obvious — and provided you are a determined tinkerer with time on your hands. Some TAC models are contained in pressure-molded plastic cases that must be (gently) broken open to gain access to the electromechnical innards, including the end of the handset cord. This also means a delicate gluing job to put it all back together properly.

More expensive instruments are often worth fixing. Whether to try to undertake the repair yourself or take the broken phone into a regular phone repair shop depends on the nature of the problem as well as your skills and tools, not to mention the availability of any needed replacement parts.

Troubleshooting a phone problem is more likely to lead you to a repair shop than to a tinkerer's workbench. But troubleshooting is often a necessary step in locating the exact piece of the equipment that needs repair and getting the right parts to make the repair.

Phone Company Problems. There isn't anything you can do about telephone problems you can trace back to the phone company's end of the sýstem — except to inform (and convince) them the problem is theirs and wait for them to fix it.

Isolating Problems You Can Fix. The telephone wires and cords and the telephone instruments in your home or apartment, however, are your responsibility. Before doing any repair, your first step is to diagnose the actual problem correctly and isolate the faulty part of your phone system to either a specific telephone instrument or to its cord, or to the wiring connecting the jacks and junctions in your system.

Be Sure There Is a Problem. The first troubleshooting step is to be sure there is a problem in the equipment, not just "human error" in operating it. In many cases, when phones don't work, a quick survey of all the locations in your system will reveal a phone left off the hook. This will happen when someone lays down a phone to pick up an extension somewhere else, then after a long conversation forgets to go back and hang it up. It can also happen with some of the disconnect switches (especially on TACs) where the phone has not been seated properly in the cradle or placed securely on a flat surface to disengage the phone.

Be sure the party you've called last has hung up. In some cases, a phone company won't disconnect either party until *both* have hung up. Others will disconnect only when the calling party has hung up. Most will disconnect when either party hangs up.

Get a Good Phone. For almost all troubleshooting work, you'll need a phone you know is in good working order. Take your phone to a neighbor and see if it works on that line. If you

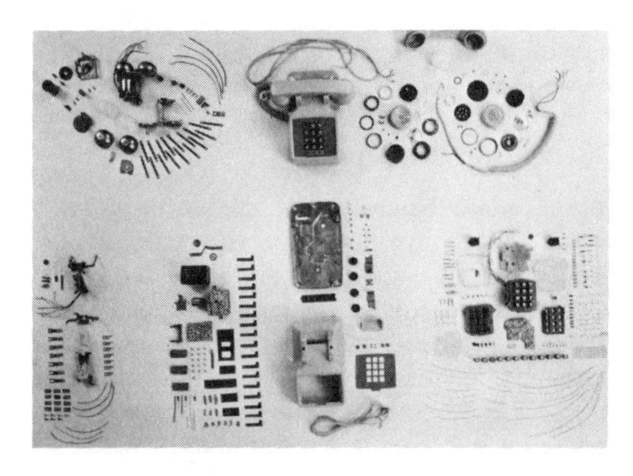

Many of the parts in today's phones are not built to be repaired, but to be replaced as a modular unit.

Many apparent phone problems can be quickly traced to a phone left off the hook or hung up improperly.

One-piece phones like this often do not disconnect because they are set down on clutter or their own cords.

can't check out the working condition of any phone in your system, and can't borrow one from a neighbor, you may have to make an investment in a very inexpensive TAC to do the testing.

Isolate the Problem. Try to isolate where the problem is by disconnecting all the phones in your system one at a time, to see if the problem is corrected with any one unit disconnected. If so, you've isolated the problem to one specific phone or cord. If the problem persists, even with extensions disconnected, you'll have to venture further into the process.

Substitute Phones. Start at the modular jack where you first noticed something wrong. Unplug that phone and substitute one you know is in good working order. Connect the good phone to the cord where it connects to the back of the phone, not at the wall. If the problem clears up, you have isolated your problem to the phone that was there. If the problem persists even with a phone you know is in working order, you've isolated the problem to the cord or wiring leading to that location.

Substitute Cords. Now use the modular connecting cord that was with the phone you know to be in good working order to connect the phone to the wall jack. If the phone works with the new cord, inspect the old cord for damage or loose wires at the connectors. You'll have to replace or repair the cord no matter where the damage to it is.

Bypass Wiring. If the problem persists even with the new cord, the trouble is probably in the wiring between modular jacks. Try to isolate where in the wiring the problem is. Start at your end of the phone company's terminal in your home. Some of the telephone company's terminating points do have modular plugs to connect to the rest of the house. Try your good phone there. If you still don't get anything, the problem is most likely with the phone company's lines.

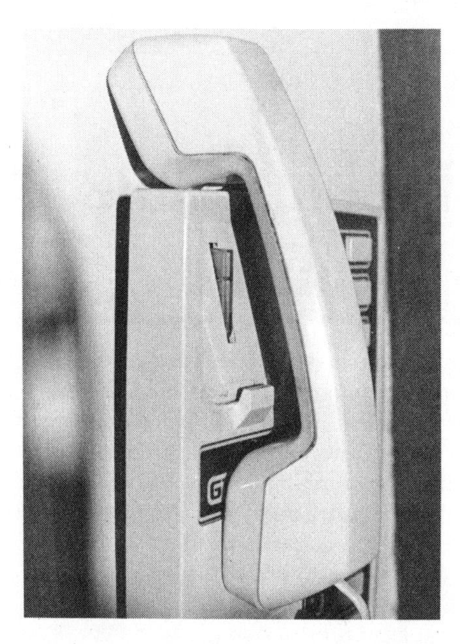

Even a wall phone can be accidentally left "off-hook" if not seated properly in its cradle or catch.

If the phone company's lines coming into your house are okay, move to the first wire junction or modular jack along your wire route, disconnecting the wires that go to the rest of the system, and trying your good phone on the line. If that location is good, hook up the wires again and move to the next location along the line. Do this until you have isolated the problem between two locations in your system.

Coping with the Repair

If you isolate the problem to the wiring and not to the phone, turn to Chapter 18, where repair procedures for your wiring are discussed.

If you have isolated the problem to a specific phone instrument, the troubleshooting procedures in Chapter 17 will help you pinpoint the problem and influence your decision about fixing it yourself, having it repaired or investing in a new phone.

17 Troubleshooting Telephones

The telephone instrument's basic functions of translating sound into electrical impulses and back again are done with relatively simple, straightforward electrical circuits and parts. When the phones are asked to do more than this, troubleshooting and repair problems become more complicated. The more sophisticated the function, the more sophisticated the electrical and electronic circuitry controlling it. But no matter how sophisticated, an electronic part or an electrical path for current can be diagnosed and repaired.

Seen and Unseen Sources of Problems

The integrated circuits in most modern telephones can be very easily damaged by voltage surges or static electricity. In tests of new telephones, Consumers Union subjected phones to an 800-volt surge of electricity, like the kind that might happen in your home during an electrical storm. This rather brutal test left some of the phones completely blown, while others wouldn't ring or dial correctly. Some survived. Static electricity (the kind you can generate by scuffing your feet on a wool rug when the humidity is low) can quickly destroy a microprocessor. Consumers Union zapped each phone they tested with an 8000-volt burst of static electricity and found this less of a cause of phone malfunction than voltage surges, unless the zap was in just the right (or wrong) place.

Today's sophisticated phones have sensitive electronic components that are hard to repair.

Warranty Repairs

Most phones can be repaired or replaced under warranty at the store where they were bought. It's best to check this policy with the store before you buy. Sometimes the stores themselves don't do the work, but have arrangements with local "authorized service centers" somewhere near where you live. Sometimes you have to send it back to the manufacturer. Fortunately, most are on this continent even though phones are made elsewhere. Better check, however, before you buy and find out you have to pack up a phone and send it to Taiwan for repairs. Many phone equipment companies are importers, not manufacturers. Most, however, do have arrangements with authorized service centers to stock spare parts and do repairs on the phones they sell.

If you didn't buy your phone from a manufacturer with a local service outlet, you will probably have to send the broken phone away to a factory-designated repair facility listed in the warranty documentation or in the owner's manual. Be sure to read and follow the exact instructions for handling such a situation. In some instances, the manufacturer must authorize you to ship the phone before the repair service will do the work. You may also have to use prescribed shipping cartons. Whenever you invest in a new telephone, make an additional investment in some storage space in your basement, attic or garage to keep the carton and packing material the phone came in — and the warranty information so you know how, where and if you can return the phone for repair.

How to Recognize Symptoms of Phone Problems

There are only a few things that can go wrong with a phone, and even fewer you can fix yourself. Symptoms of phone problems are usually

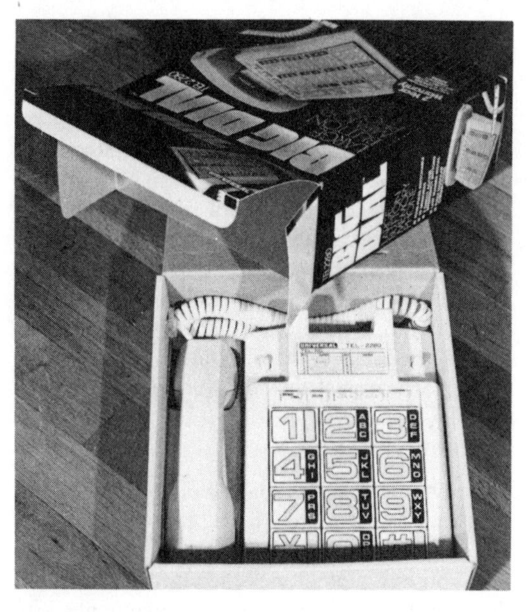

The best material to pack a phone for shipping is the packing and box that it came in when you bought it.

When shipping a phone, use a sturdy carton and protect it from damage with foam packing.

limited to (1) a completely dead phone, (2) no tone for dialing, (3) phone does not ring, (4) you can hear people but they can't hear you, (5) people can hear you but you can't hear them, (6) noise on the line and (7) RFI.

Phone Is Completely Dead

Symptoms. If you pick up the phone and can't hear anything you've got a problem. Sometimes, you can hear the sound of your voice in the speaker of the phone, but can't get a dial tone. These are usually symptoms of something wrong with your phone, although the problem can be in the wiring.

Troubleshooting. First, see if you are still *getting* calls on your phone line with the proper ring and voice connection. Try another phone that you know is working for this test. Then, ask a neighbor or friend to call your number. If your phone doesn't ring when they call, the trouble may be in the phone company's line or in your wiring. If you are, however, able to receive calls but can't get a dial tone to dial out, you'll have to isolate the problem to see if it is just one phone or your entire wiring system.

The first thing to check if you don't get a dial tone is the other phones in the house. Maybe someone left one off the hook and the line is still open, preventing anyone from making a call.

If you're sure it isn't something as simple as that, switch with a phone that you know works. Again, your neighbors may have to help you out by "lending" you their lines to check out a phone. Try other outlets in your home system first. If your phone still does not bring in the exchange-provided dial tone, you know it's not the phone instrument that's causing the problem. It must be in the wiring that leads from or to the troubled jack. It could also be in the cord between the jack and the phone itself.

If you detect that the phone instrument is in fact the source of a dead line, you should check out the coiled cord linking the handset and the base. This is a frequent source of phone trouble. Those cords are likely to receive a lot of pulling, twisting and other wear and tear. Substitute the handset cords, but be certain you're plugging in one known to be good. This cord test is a cinch to do if you own a high-quality instrument. These non-TAC phones invariably come with a modular coiled cord that can be quickly unsnapped at each end. To conduct your test, you must have a spare cable having the correct jacks at both ends. It need not be coiled. Many people find it wise to keep a spare handset cord handy, anyway.

If the troublesome extension now works, you can bet the malfunction was caused by the original cord. If not, the problem is likely to lie in the wall jack, the wiring leading to the main terminal or in that terminal itself. Where possible, use a spare modular cord and plug to connect a good phone directly to each of the modular jacks in your system to see which of them is actually working. When you find one that doesn't work, you've isolated the problem to the wiring between two jacks.

Repair. If none of your phones has the exchange-supplied dial tone on the line, or if all of them are dead and you've eliminated both the phones and your wiring as the source, the problem is likely to be in the telephone company's system outside your home. Go back to your neighbor's house, or to a pay phone, and call the local phone company to report the difficulty. Incidentally, you may have to convince them that the line fault is really theirs. Be ready to cite your evidence.

If you've pinned down the problem wiring, you can plan how to make a replacement. This is an effort that may not always be easy. You may be forced to pull out the old wiring and install a new system in part or all of it. We'll give you more help on this task in Chapter 18.

No Tone for Dialing

Symptoms. If you cannot dial out on your tone phone or consistently get wrong numbers, something may be wrong with the electronic circuits that generate the dial tones. This same symptom can be found on rotary dial phones when the dial does not spin back smoothly or evenly.

Troubleshooting. Even if you can't hear the electronic dial tones in the speaker of the phone, try dialing the number of a friend or neighbor. If they answer, the problem's not in your dial but somewhere else in your phone. Don't try some of the recorded messages services like time or weather for this test of how well your phone is dialing. Many of these have sequential answering equipment. With this device, the service may have a dozen or more lines coming in (while they list only one number in the directory) and when one is busy, the equipment automatically switches to the next number in their sequence that is not in use. If you dial a number like this, you may be sending out the wrong number with your dial but if it's still one of the numbers in the service's sequence, it will answer and you will still think you've dialed the right number.

Switch contacts in some of the cheaper TACs may be sending only one of the dual tones needed for DTMF dialing. You can spot this condition by listening to the sound of the tones carefully as you press the button on the keypad.

Handset cords are built to withstand only so much damage and abuse before they break down and need repair. This may often be the source of problems you think are in your phone.

Repair. If you own a genuine tone phone, you will want to avoid attempting your own repairs on a correctly diagnosed dialing malfunction. The tone generator system built into the instrument is a miniature (but narrowly specialized) computer. Its circuit system is sophisticated and some required replacement elements are extraordinarily hard, if not impossible, to find or buy on the retail market.

Tone-generating dial systems are far too varied from model to model for us to discuss in any helpful way. Tinkering with these circuits can also result in other problems for you and for your local phone company. If you do the repair improperly, you may find that dialing is either impossible or even more inaccurate. This will almost always mean your dial-outs ring the wrong lines.

It is best to leave repair of the tone generator circuit systems to a properly trained and equipped technician.

Phone Does Not Ring

Symptom. Sometimes your phone works okay except it just does not ring. This unfortunate symptom must usually be brought to your attention by somebody outside your household who has tried to reach you when you were known to be home.

Troubleshooting. First, check to be sure nobody has switched off the bell or the ringer.

Many new phones will let you do this with a switch mounted on the base section. On poorly designed phones, the ringer switch can be thrown by accidentally touching it when you pick up the phone by its base.

If you have only one phone with the ringer switched on, or if other phones in the system ring, you're probably safe in assuming the ringer is broken. Arrange for a friend to give your line a call to verify the problem.

Repair. If you have two or more phones, you may be faced with the problem of having a REN total of more than 5. The REN is (or should be) listed on the phone with the FCC certification notice. When hooking extensions into your system, you must add up the RENs of all the phones. If the total exceeds the number 5, the voltage and current supplied by the phone company may not be enough to ring your phones. At best, a REN total higher than 5 will give you intermittent ringing service. At worst, it will give the phone company problems and they can disconnect your service until you fix the situation.

If you've established that the ringer (or bell) in one phone just isn't working, you can unplug the phone and carefully remove the cover from the base section. Look for the bell. Notice if there are any obvious loose wires leading to it. Find the bell clapper to see if it has become stuck.

If you can see where to reconnect a loose wire, do so. If it's the clapper, gently unstick it and see if it moves freely. Put the case back on and arrange for someone to call your number. If there's a ring, you've succeeded in fixing the problem. If not, better take the phone to the shop for repair.

Many phones on the market today do not use conventional bells. They contain electronic ringers with the warbles and chirps so widely associated now with newer phones. If one of these malfunctions, you throw it away and attach a new one. But, there's a big caution here. The replacement ringer must be the exact electrical equivalent of the original (that REN matter again). If it isn't, it might not ring at all and might even affect the operation of your other phones as well. Best to follow the rules.

If you don't have a proper replacement for a ringer or a bell, take the phone to an appropriate repair shop. Maybe — just maybe — they'll sell you a replacement part.

Avoid attempting repairs on the electronic printed circuits of a tone dial pad. They are precisely tuned circuits that damage easily by tampering.

You Can Hear People; They Can't Hear You

Symptoms. If you can hear other people talking to you on your phone, but they can't hear you, the condition is very likely to trace back to a broken microphone in the phone you are using or to the wiring and cords involved. It is often an intermittent problem, happening only now and again.

Troubleshooting. Through a process of elimination and substitution, isolate the problem to the telephone, not to the cords or wiring. (These troubleshooting procedures for cords and wiring are described in Chapter 18.)

If the problem is in the telephone mike rather than wiring or connectors, you face a choice. Look inside the case or take it immediately to the shop.

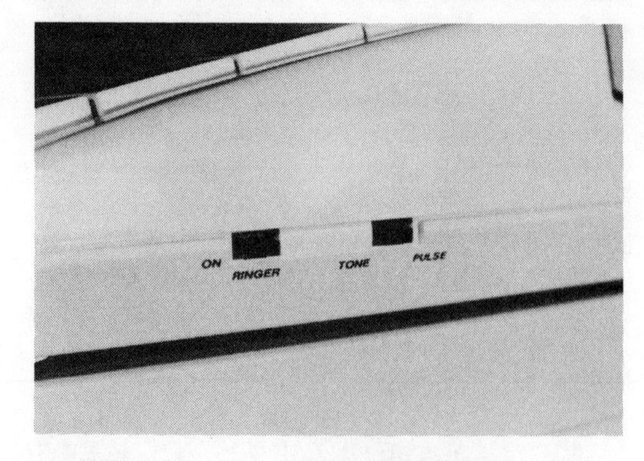

On a phone like this, the ringer can be accidently turned off when you think you are switching from tone to pulse dialing. You may think you have a defective phone if it no longer rings when someone calls, but all that has happened is the ringer is turned off.

If you own a better-than-TAC phone, you might be able to screw off the case element covering the mike. The mike is in the end of the handset you speak into. You may have to gently lift out the mike to see if any connecting wires have come loose. If not, or if you simply can't see any sign of physical damage, then you're far better off taking the phone into the shop.

Repair. First, if you have a phone with a carbon granule mike — and many older phones do — gently slap the phone in the palm of your hands once or twice to "unpack" the carbon granules. This just might solve your problem.

A proper replacement for any carbon, electret or dynamic mike is not likely to be carried as a spare part by phone stores. Connecting it into the case may well involve fittings and procedures peculiar to your phone model. It's virtually impossible to give universal guidance here.

People Can Hear You; You Can't Hear Them

Symptom. When people can hear you, but you can't hear them, this is very likely to be the symptom of a broken telephone speaker inside the handset. If you do, however, hear your own voice through the speaker but still don't hear other people who call, the problem is not in the speaker and you should apply troubleshooting procedures for other problems.

Check for loose wires or a stuck clapper when a phone with a bell ringer does not seem to be working.

Check for loose or damaged wires on the speaker in the earpiece of phones where you can get at it easily.

Troubleshooting. This condition often takes the form of a malfunction that happens only when you hold the handset in a certain way. This strongly suggests a loose wire leading to the speaker — one that loses its electrical contact when the handset is tilted in a particular position.

Repair. Treat this problem as you would a broken mike — and with the same likelihood that you would probably be wise to get the phone into the repair shop for proper replacement of the malfunctioning part. Here again, a spare part having the correct impedance and other electromechanical characteristics for your particular phone model may often be difficult to come by.

Noise on the Line

Symptoms. Static, scratches, garbling or howls on your line do not necessarily indicate that the phone company's exchange is having a problem. It may be the result of a breakdown inside your own system.

Troubleshooting. Try to isolate the problem to a specific connector or stretch of wire following the troubleshooting procedures described in Chapter 18. A broken ground connection somewhere in the system is a likely culprit.

You may also find the noise is entering the system because a wire or cable has lost a significant section of its insulation. Electrical wires or speaker wires to your stereo system can induce hum or noise into your telephone wires if they are too close. Telephone wires have been confused with cables associated with stereo systems or microcomputers, resulting in

On some phones you can unscrew the mouthpiece to check for loose or damaged wires.

incorrect hookups. This situation will usually produce more than just noise on your line! (Depending on what you've mistakenly hooked into, you may also hear a lot of noise from your phone company's office about it.)

Repair. One particular garbling noise on the line can be caused by too many phones off-hook at the same time.

Hum can be caused by phone wires running too close to AC power wires or stereo speaker wires. Try moving the wires farther apart. The electrical field causing the hum decreases by the square of the distance from it, so you don't have to move the wires far to get a noticeable difference. Wires that cross at right angles will produce less hum than wires that run parallel to each other for some distance.

Go through your system checking all modular jacks, mounting plates and wire junctions. Be sure all screws connecting the wires are tight and that there are no wires shorting out two of the terminals. Be sure the yellow ground wire is correctly connected in each of your jacks.

If replacing a cable or tightening ground and other connections in your modular hardware does not clear up the noise, a call to the phone company may well be in order.

Radio Frequency Interference

Radio frequency interference (RFI) can definitely invade your telephone, although you may be fortunate and never have the problem. Today's high-tech phones with their solid state devices are more prone to RFI than older phones were. RFI can come from many sources: commercial radio stations, amateur radio, CB radio, microwave, even your home computer.

Symptoms. Radio frequency interference may or may not be heard as intelligible sounds on your phone. Many times interference simply shows up as a loud hum or the volume of the person's voice you're talking with suddenly begins to fade in and out. A single-sideband signal (transmitted by amateur radio operators, ship-to-shore radio and some FM stations) will be garbled badly no matter how loud it is. A strong FM signal will usually cause just a loud hum. On the other hand, an amplitude-modulated signal, such as an AM broadcast station or most CB radios, will usually be clear and readable.

Troubleshooting. RFI can be picked up by the telephone company's wires and sent to your home telephone system, or it can be picked up directly in your home telephone system. When the phone company's wires are up in the air on poles, they constitute an antenna that runs all over the neighborhood and an obvious target to receive RFI. Underground telephone wires are less prone to receive this interference, but they can pick it up all the same.

Direct radiation inside your home is easily recognized because it will change in intensity as you move your phone around the room. Fortunately, this type of interference is rare.

Repair. Getting rid of RFI coming into your home from the wires is comparatively easy compared to stopping it when it invades your home telephone system. Once you have determined that you have RFI, call the local telephone company. If the RFI is being picked up by their lines before it gets to your house, they can install a pair of bypass capacitors to ground from each of the telephone wires leading to your house. Do not attempt to do this yourself. This is the phone company's responsibility and there should be no charge to you.

If the problem can be traced to the phones or circuits inside your system, you can try "choking" the high-frequency RFI with a pair of inexpensive RF chokes, available at most radio supply stores. You can use either 10 or 100 μH (microhenry) values. You will need two of them. The 10 μH choke is smaller and probably easier to work with because space is limited inside the wall jack where you install the chokes.

You put the chokes in series with the red and green wires of your phone line at each phone

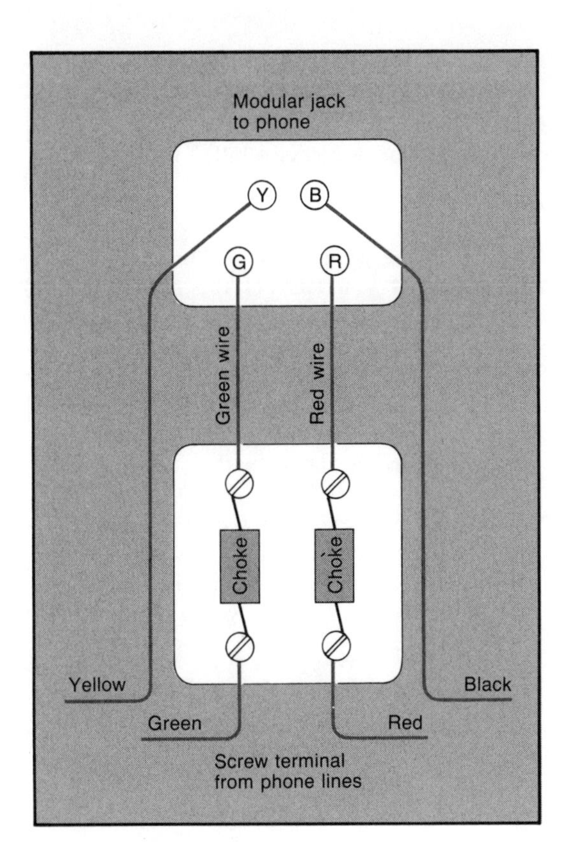

where you notice interference. You can mount them in a separate four-screw terminal, connecting each choke across two of the screw terminals and then connecting the red and green wires of your phone line to one side of the chokes and the red and green wires to the modular jack for your phone to the other side. Be sure to extend the yellow ground wire to the modular jack without going through the chokes.

All chokes have some resistance, and if your phone system's REN capacity is marginal, the extra resistance could cause the phones not to ring.

If this does not solve your RFI problems, purchase a pair of inexpensive torroid cores from a radio or electronics supply store. These cores look like miniature doughnuts and are made of powdered iron mixed with glue. You may need to experiment with the inductance values of these cores. Remember when you buy them they'll have to fit inside the case of your phone. You might even take your RFI-plagued phone with you when you go shopping.

Open the telephone cradle base unit and loosen the pigtails connecting the components inside your phone to both the handset jack and the line cord jack.

Pigtails have four leads (red, green, yellow and black) and are connected with push-on clips to your phone. Wrap the red and green pigtail two or three turns around a torroid core and then reconnect the pigtails to the phone with the push-on connectors. The inductance of these cores forms two RF chokes, keeping all RF from passing along the cords into the phone.

By the way, these torroid cores will clear up this same problem if it exists with your stereo sound system. Simply wrap a few turns of your speaker leads around the cores at the point where the leads connect into your stereo.

You may also have to experiment a little with where you put the torroid cores in your phone. There is the danger of creating a tuned circuit between the base of the phone and the coil around the core that will amplify the signal causing the RFI, not suppress it.

Most true phone stores sell modular RFI filters that you can install in the line between a wall jack and your phone. This relatively inexpensive accessory ($10 to $15) will save you the trouble of taking your phone apart and experimenting with the values of torroid cores or RFI chokes. These ready-made RFI filters are built with chokes and torroid cores that have the correct values to suppress most AM and FM radio frequency interference commonly found to plague home telephone systems.

If the RFI is being picked up by the printed circuit boards inside your phone, you can minimize the interference by turning the phone so the edge of the circuit board points toward the RFI source. More of the interference signal will be picked up when the components mounted on the flat side of the board face the source than when the surface area facing the RFI is minimized as it is on the edge of the board. This is much easier to do, of course, when the printed circuit boards are in the handset than when they are in the base of the phone. It might prove to be a little difficult to hang up your cradle phone when the base is on its side.

General Repair Tips

Don't Tinker with Phones. Make repairs only when you know what you're doing and the malfunction is obvious. Experimentation has no place in telephone repair when the results of failure can lead to damage of phone company equipment, affecting other people's ability to communicate on the system.

Work from Easy to Difficult. When you start

your troubleshooting, begin looking at those problems that will be the easiest to fix. With luck, your search will stop with one of these parts of the system. If it doesn't, work progressively toward the more difficult repair. There's no sense in trying to tackle a hard job when you can do an easy one with the same results.

Work from Cheap to Expensive. Test out those parts that will cost you less to repair or replace. A broken connecting cord costs only a few dollars; a replacement phone is likely to be a little more expensive. Check out the cheapest repair first and hope you've located the problem. Again, there's no sense in spending more on the job if you can get by with an effective inexpensive repair.

Don't Call the Phone Company with Every Problem. Most phone companies do offer repair services but not many will even look at equipment you purchased somewhere else. If they do make a house call, be prepared to fork up a tidy sum to pay for the service bill.

Document the Symptoms of Phone Company Problems. If you suspect the problem is in equipment the phone company is responsible for, be sure you have gathered all of the evidence to support your allegation before you call. You'll be asked to prove the problem is theirs, not yours. If the phone company suspects you have a phone that's causing trouble on the lines, they'll ask you to disconnect it. If the problem still exists they can — and will — cut off your phone service. If this happens you may have a bad phone, problems in your wiring or you may have made the wrong repair.

18 Checking Wires and Connectors

Troubleshooting your home telephone system will often isolate a problem to your wiring or to the jacks, terminals or connectors you have installed. Simple repairs can often save costly replacement. Where repairs are not possible, you should isolate the problem to a specific part. This will save both time and cost when replacement is needed.

What Can Go Wrong?

There is not much that can go wrong with a length of insulated copper wire or a box with screw terminals in it. Problems in the wiring or connectors are usually limited to:

- mixed-up color-coded wires
- broken or damaged cords
- broken or damaged wires
- shorted wires or connections
- loose connections
- missing connections

Mixed-up Color-Coded Wires. A common cause of phone problems is mixed-up color-coded wires. Look at the bottom diagram. As you look into the hole for the plug on a wall jack (with the notch for the spring clip up), the yellow wire terminates in the gold-strip contact on the left, followed by green, red and then black on the right. In the jack on the telephone the order is reversed. With the notch for the plug's spring clip up, the black wire is connected in the jack to the contact on the left, followed by red, green and then yellow on the right.

When you make or repair modular cords, this is why it is important to install plugs with the yellow on the left on one end and the yellow on the right side of the plug on the other end of the cord. While logic may tell you to cross-over the wires so yellow is always on the left looking at the jack, the phone (as the top diagram on the next page shows) expects the wiring to come straight through with the yellow on the right, followed by green and then red and black. If the red and green wires are switched, the phone will not work.

Broken wires or cords are caused by rough treatment and prolonged exposure to normal household hazards.

This may be a source of phone problems in your home wiring system. The bottom diagram on the next page shows a modular telephone cord connected between two wall jacks to jump over some obstruction to the hidden wiring inside the walls. When used this way, all the wires are reversed at the second jack. The red shows up on the green terminal, green becomes red and the yellow ground is now on the black terminal.

While you may be tempted to correct this problem by making up a special modular cord (with the color coding reversed on one end), it's best to correct the color coding in the second wall jack. You may mistakenly use the cord to connect directly to a phone later, and wonder why the phone suddenly won't work.

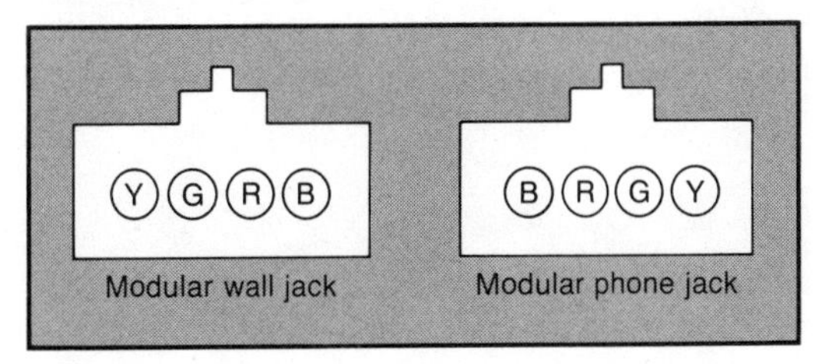

Be sure to follow directions about color coding carefully when making a new modular cord. The jack in your phone expects the yellow wire to come in on the right. Modular wall jacks expect it to be on the left.

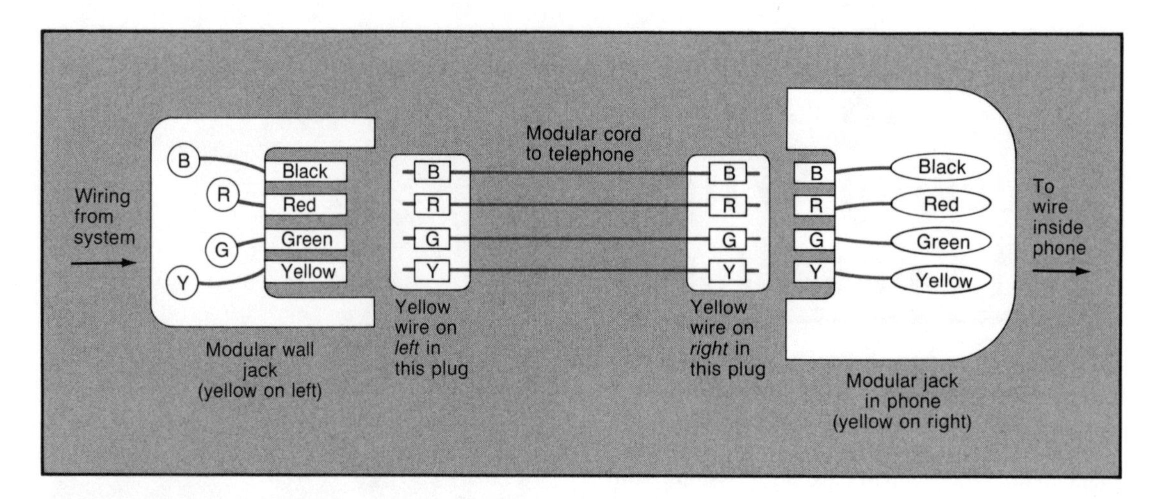

The yellow wire must be on the left on one end of a modular cord and on the right in the connector at the other end. Follow this color coding carefully or the red and green wires are crossed and the phone won't work.

Broken or Damaged Cords. Most of the time, broken wires or cords are caused by rough treatment when someone uses the phone. Extension cords, openly exposed to traffic patterns and vacuum cleaners, are especially prone to damage. Coiled handset cords, even though specifically designed for rough treatment, are gradually destroyed when they are twisted or stretched too far.

Telephone cord replacement is simple and relatively inexpensive. Packaged *telephone cord* is available in two ways: with a modular plug on one end and a spade lug on the other (for the screw connection to the old 42A wall block), or with modular plugs on both ends. Unpackaged telephone cord is also sold in whatever length you request while coiled *handset cord* is sold in standard lengths up to 25 feet.

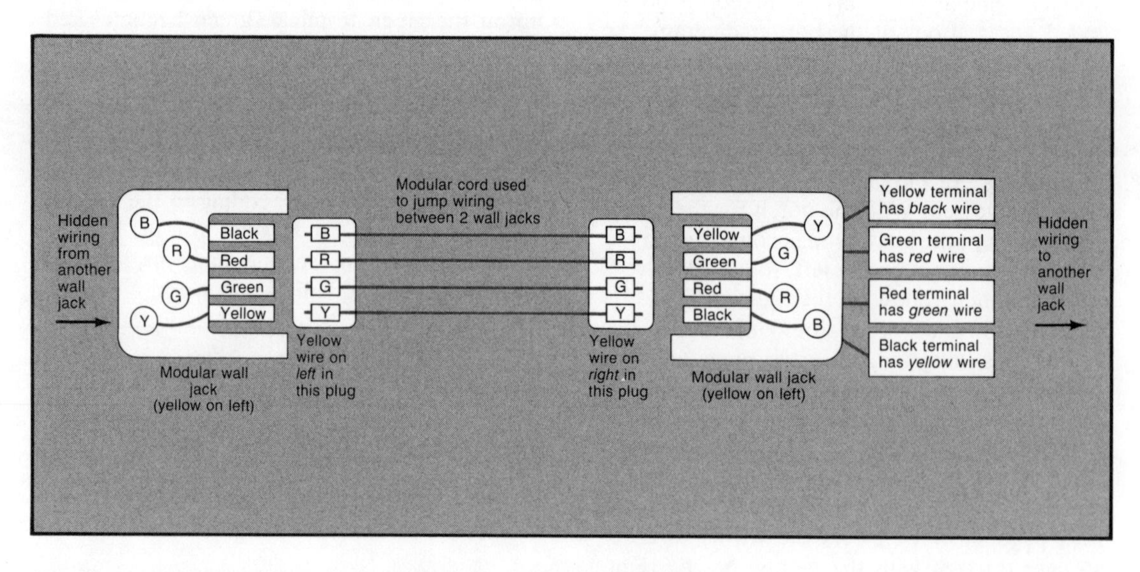

When you use a modular cord between two wall jacks, as you would to jump over an obstruction to in-the-wall wiring, the connections at one jack will have to be reversed to make the color coding agree with the rest of the system.

One caveat about shopping for the clear plastic modular plugs is to specify whether they are handset cord plugs or telephone cord plugs. Both plugs contain the identical four wires, but the telephone cord plug is six terminals wide, while the handset cord plug is only four terminals wide. They are not interchangeable.

Replacing broken plugs or making new modular cords is almost impossible to do correctly without a special modular crimping tool. This tool looks like an electrician's wire stripper. Some modular crimping tools will work only on the wider telephone cord plugs and cannot be used on the narrower handset cord plugs.

Installing a modular plug on a telephone cord only takes a couple of minutes. First, you cut the cord to the desired length using a pair of diagonal side cutters. Sometime the tool will have a cutter for a neat, straight cut. Make the cut as close to a right angle to the cord as possible. Next, use the stripper section of the tool to remove the outer insulation, being careful not to break the color-coded insulation on the individual wires inside. Metal "teeth" inside the modular plug pierce this insulation to make contact with the wire when you crimp the plug on the wire. Removing the outer insulation involves inserting the wire until it hits the stop and then closing the tool to remove the correct amount of insulation.

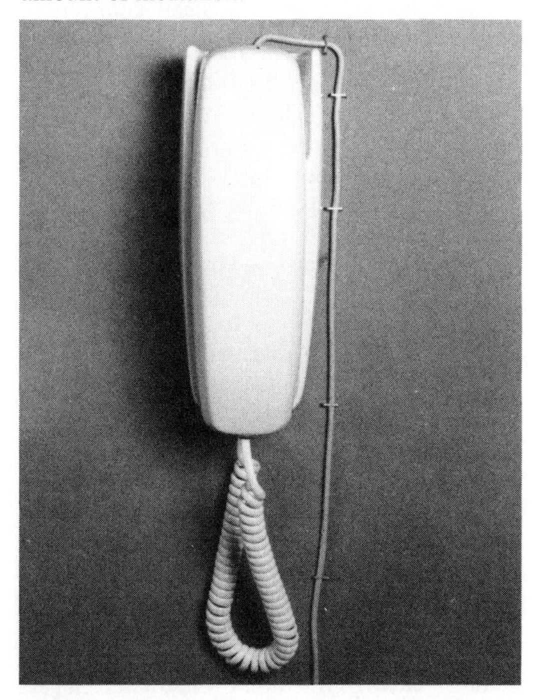

Failure to use the correct kind of staples to fasten telephone cable can easily result in a broken wire.

Hold the modular plug with the spring clip facing you and insert the cord from the back with the yellow wire on your left. Be sure the cord is inserted far enough for the ends of the conductors to be flush with the tip of the plug. Insert the plug and cord into the crimping tool,

The most common place to find shorted wires is in a junction or wall jack. Here, the end of one wire has not been trimmed and is touching another screw terminal.

The two terminal wires in this jack are shorting out. This can happen when the screws are tightened down too tightly and the spade lugs twist together and touch.

and press down until you reach the stop. On the other end of the cord, the yellow wire should be on your right as you look down on the top of the modular plug. Fasten this plug in the same way you did the first. This correct matching up of wires is important.

If the modular crimping tool does not have a stop to control how much pressure you put on the modular plug when you install the wires, you have to be careful not to press so hard you break the plug; you won't hear a click. Easy does it!

Broken or Damaged Wires. The four-conductor, color-coded wires in your system are often damaged by rubbing against rough surfaces, outside or inside your walls. Wire can also be quickly damaged when you try to staple it to your woodwork or walls with a staple gun. Another common cause of cable damage occurs when you've made too sharp a bend around a door or window frame. Another place wire can break from being bent too sharply is in the modular jack boxes. The solid copper wire in most home telephone cable cannot take the same kind of bend that the stranded wire in an extension cord can.

Shorted Wires or Connections. The most common place to find shorted wires is in a terminal box or wall jack. They're the only places in your home wiring system where bare wires are allowed. Often the loop of wire around the terminal screw is too long and touches the other contacts. Pieces of cut wire can also find their way between two contacts.

You can't tighten a terminal screw completely if you try to put too many wires on it. This can result in an intermittent connection and can be difficult to trace.

Another kind of short is the *capacitance short*. This happens when damaged insulating material is thin enough to act as a dielectric, letting small amounts of current leak across, even though there is not a direct metal-to-metal short.

Air can become a dielectric, giving you a *capacitance contact*. If the wires are a little loose on the connectors, expansion or stress can separate the wire from the connector just enough to create noise or lowered volume on your phones, but not enough to open the contact completely.

Loose Connections. Having too many wires on one terminal screw often prevents you from tightening the screw completely. Any kind of movement or stress on the wire eventually finds its way back to the terminal screw and can gradually loosen it.

A broken latch on a modular plug can also be the source of a loose connection. This latch not only keeps the plug from being pulled out accidentally, its spring action helps to make the contacts between the very small connections more secure.

Missing Connections. The most common way a missing connection happens is when someone forgets to plug in a phone or forgets to replace a borrowed extension cord.

The connection between the telephone company's terminal and your system's terminal is another source of missing connections. In condos and apartment complexes you may have connected your system to the terminal the builder installed but find that no one assumed the responsibility of connecting that terminal to the one the phone company installed.

Before You Begin

Before you begin any troubleshooting procedure with your telephone system, disconnect your system at the phone company's terminal box. Often this is done by just unplugging a modular jack. While telephone voice current and voltage is usually not harmful (or even felt in most cases), you don't want your fingers crossing two of the wires when your phones ring. While the current of this high ringer volt-

Failure to follow the color coding on your connections will almost certainly result in problems with your phones, and may cause problems with the phone company's line.

The ends trimmed off wires in a junction or jack will eventually find their way across two wires or terminals where you never intended a connection to be.

age is small, you'll still feel quite a tingle — or more.

Begin your troubleshooting with the easiest problem and the cheapest solution first. There's no sense in doing the difficult job or making the expensive repair if you can solve the problem more easily and cheaply.

Checking Connections

Do a visual inspection of your terminals, outlet and connectors before you take on the more difficult task of checking all the wiring between connectors.

Check Color-Coded Connections. Check first to be sure you have the proper color coding on your connections. Be sure the yellow, red, green and black wires are correctly paired on the same terminal screw. In some jacks, color-coded wires lead from the screws to the jack connectors in the unit. On others, the connections are internal and the screws will be marked either by a color coding on the case or by **R, G, Y** or **B** stamped on or near the screws.

Tighten Screws. Remove the covers from the connector box or terminal and tighten all the screws, even though they look tight to you. Sometimes just a quarter turn will fix your problem. Screws should be snug — you don't have to torque them with heavy-duty tools. Most of the connectors screw into the plastic base, which is easily stripped if too much pressure is applied. If this happens, you'll have to replace the entire unit.

Check for Shorts. Check all the connections to be sure that no wire "tail" extending from the loop around the screw is making contact with another screw or with the bared end of another wire. Look inside the unit carefully to see if the piece of wire you have snipped off the end has become lodged between two contacts or bare wires. You may have to remove the connector from the wall to shake these wayward pieces of wire out.

It is possible — but highly unlikely — that you bought a defective modular jack or terminal with a short you can't see inside the plastic casing. You'll have to use an ohmmeter to check for continuity (full needle deflection) between each of the screw connectors after first disconnecting the wires.

Checking Wires and Cords

Troubleshooting wires and cords begins with a visual inspection and progresses to continuity testing and finally replacement testing.

Visual Inspection. Look at all your wiring in an orderly, systematic way. Start at the telephone company's terminal and trace each wire to its terminal or outlet box. Look for evidence of damage in missing, frayed or melted insulation on the wires. Check especially where vacuum cleaners or traffic patterns could have run into the lines. If you've stapled the wires to any surface, check each staple. Be sure staples do not pinch the wire or break the insulation. Staples should be loose enough to let the wire slip a little, but tight enough to keep it in place neatly.

Look at the plugs of your phone-to-wall cords. Constant use and abuse could loosen the four tiny wires inside the modular plug. Grip the plug and wire and pull while you gently bend the wire away from the plug. Any movement most likely means a bad connection in the plug. Check the operation of the plastic spring latch on the modular plug. If it doesn't snap back when you depress it, it probably has lost its holding ability to ensure a good contact.

Hidden damage to wires can happen when the cable is installed through walls and snags on a rough piece of wood or sharp part of a nail you've accidentally drilled through. Look at the insulation on the end you've pulled through the wall. If there are scratches or rough spots on that end of the cable, you may have more serious problems inside the wall. A continuity or voltage check will save you the work of pulling the wire out of the wall for a visual inspection.

Replacement Testing. The next check to make of any problem location in your wiring is to substitute a known good length of wire or cable. Begin troubleshooting with the cords that run from your phone to the modular jacks. They're the easiest to do. Find (or buy) an extension cord that you know works. You may have to enlist the help of a neighbor or friend and plug your cord into their working phone system to be sure your substitute is a good cord. Unplug all the phones in your system, then using the good cord connect each to a modular jack, one by one, until you isolate the cord giving you problems.

This check assumes your phone is good — but it also assumes the handset cord is good. This may not always be a valid assumption. If all the extension cords check out okay, try substituting handset cords to see if you can isolate a bad one.

If the problem seems to be in the wiring in the walls or along your baseboards your task becomes a little more difficult — and expensive. Get a new length of wire long enough to run between two jacks farthest apart in your system. Connect a modular jack or wire terminal to each end of this good wire. Then, at each place where you can get at the wires, disconnect the existing wires and connect your substitute, being sure all connections are properly color-coded and tight. If the phone works, you've isolated your problem to the wiring between those two connectors.

Continuity Testing. You can test for broken wires in your system with a volt–ohmmeter. Finding an ohmmeter with leads long enough to reach between two modular jacks 30 feet apart may be a problem. You can either run an extension wire to the meter, or check for voltage instead of resistance.

Voltage check. After all phones are unplugged — and you're sure your system is disconnected from the phone company — use a modular extension cord with one end removed and the wires bared to put a flashlight battery across the red and green wires. Be sure to disconnect *all* phones and especially the line to the phone company. Even the small voltage of a flashlight battery with the wrong polarity can

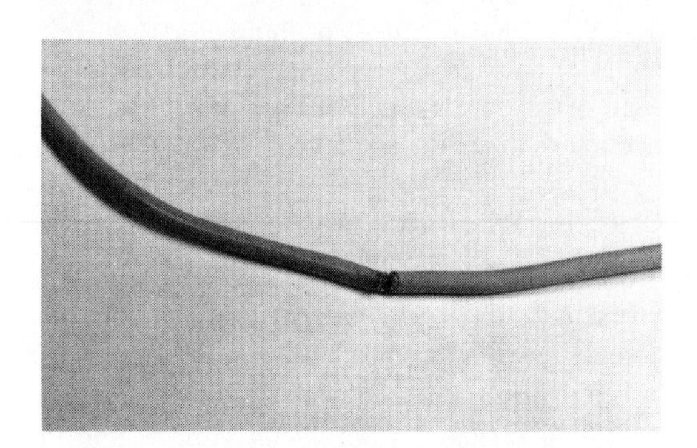

A sure indication of the source of trouble in your phone system is any kind of damage to the wiring.

Scratches or gouges in wire insulation coming out of a wall can indicate more serious hidden damage.

destroy the integrated circuits in your phone. At the other end of that run of wire, your voltmeter should read the battery voltage with the leads across the red and green wires. If it doesn't, you've got an open wire. If you get a voltage reading between red or green and the yellow wire (which should not be connected to anything now), you've got a short. Repeat the test with other pairs: red–yellow and yellow-green.

Continuity check. If you're doing a continuity check with the ohmmeter side of the meter, check not only for a reading indicating a short, but for unusually high resistance readings. Normally the length of wire alone will not offer enough resistance to cause much needle movement. If you read anything over a few ohms resistance in the wires, you've probably got a capacitance short somewhere in that section of wire. (If you read about 400 ohms resistance, you've probably left a phone hooked into the line somewhere!)

Modular plug testing. Continuity testing of modular jacks and plugs with a volt–ohmmeter can be done in much the same way, but the meter leads will be too fat to touch just one of the contacts in the tightly spaced modular connectors. The best thing to do here is to use a short extension to an extra modular jack, then use the meter leads on the screw terminals.

Repairing Knotted and Tangled Cords

Inside every coiled telephone handset cord is a snarled tangle just waiting to happen. It seems to be their destiny. The best way to cope with this problem is to keep it from happening, but that is not always possible.

Left-Handed Tangles. Almost all telephone instruments are built for right-handed people. The handset cords usually come out of the left side or left rear, assuming the right-handed person will hold the phone in the left hand, keeping

With a flashlight battery connected to the red and green wires at one modular jack, you should read 1.5 volts on your meter at the next jack along the line. This method eliminates the long extension meter leads.

Other high-resistance readings can mean a capacitance short or can indicate that there is still a phone hooked up someplace on the line you're testing, as this 400-ohm reading might indicate.

Full needle deflection means a short in the wires.

No needle movement means no connection or an open.

the right hand free to dial or jot down notes. When this phone is picked up in the right hand of a left-handed person, a twist is immediately put in the cord, followed by another twist if the phone is hung up "backward." A coiled cord in this situation is doomed.

Untwisting Cords. If the coiled handset cord becomes twisted, the best way to fix it is to let the handset dangle, holding the cord at arm's length by the disconnected modular plug. This won't work too easily with 25-foot handset extensions. The handset will begin turning in the direction of the tangle, but its own momentum will make it continue to turn, twisting the cord in the other direction. You have to let it twist one way then the other until it stops.

Catch twists before they become tangles. If the coiled cord begins to show a minor twist, remove the handset and loop it around the cord in the opposite direction of the twist. Remember, tangles develop from twists at a geometric rate.

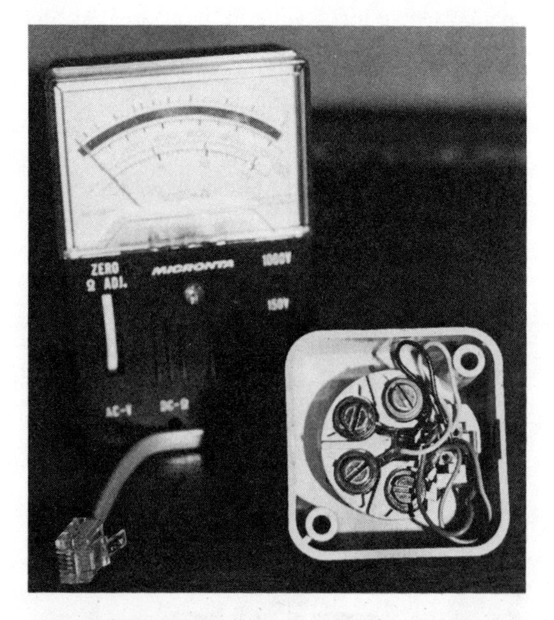

Use a spare modular jack with your meter for checking the small contacts on modular plugs. Take your meter readings off the terminal screws on the jack with the plug you're testing inserted.

Phone hung up by left-handed person already has one twist that will quickly develop into a snarl after several more phone calls.

Catch minor twists before they become snarled tangles like this. Replacement is almost necessary when cords get this bad.

Kinks like this in handset cords can be corrected if caught in their early stages.

19 Modem Maintenance

Modems come in so many models, shapes, speeds, designs and cost ranges that it's impossible to give you any sensible tips on repair procedures to use when something goes wrong.

But there are a few things you can do to isolate a problem. Sometimes, this alone will lead to a happy outcome. Occasionally, it may only help you describe the malfunction to a knowledgeable technician working in a repair shop equipped with specialized testing gear as well as with the necessary replacement parts.

There are two different situations for modem problems. One can occur the first time you try to use a new modem; the other when a familiar modem suddenly ceases to perform as you think it should.

PROBLEMS WITH NEW MODEMS

Let's look first at the new modem situation. Begin by reviewing the things you must have to make any modem communication work:

- □ a properly functioning microcomputer with which you are familiar
- □ a modem designed to be interfaced with that brand of computer
- □ communications software designed for the computer and compatible with the modem
- □ a working telephone line properly plugged into the modem

In many instances, you must also have a working telephone instrument plugged into the same telephone line and placed within easy reach of the micro's keyboard. A modular Y-adapter may be necessary to hook up both your modem and a telephone on that same line. Many modems now have that modular connection for the phone built in.

Preliminary Troubleshooting Checks

Check Instruction Manuals. The basic resources you need to install, operate and troubleshoot are the user manuals for the micro, the modem and the communications software. Failure to follow the rules laid down in the various manuals is the most likely source of your problem. But don't feel stupid. There are many peculiar rules — and the manufacturers don't always make them clear and simple.

If you're trying to dial into one of the data base services (or even a complex BBS), you may also need a manual or an instruction sheet supplied by the operator. Pay closest attention to the materials dealing with the dial-in and sign-on procedures.

Check Connections to the Power. Don't forget that all microcomputers require electric power from a wall plug or built-in batteries. Your modem also runs on an electric feed of the one sort or the other. A few modems, like modem cards in the IBM-PC or its clones, take their power directly from the micro to which they are physically attached.

Check Connections to the Computer. If you have a stand-alone modem, check the RS-232C connections between the modem and the computer. Most of the DB-25 connectors have screws on each side of the connector to hold them securely in place. See if these screws have been tightened securely. For internal modem cards, be sure the multipin edge of the board is securely seated in the socket at the end of the expansion slot. If it is slightly off-center vertically, the contact strips on the board can be making contact with the wrong connectors in the computer.

Check Connections to the Phone Line. Is the phone line plugged into the modem? Is the line active? Is the telephone on the same line con-

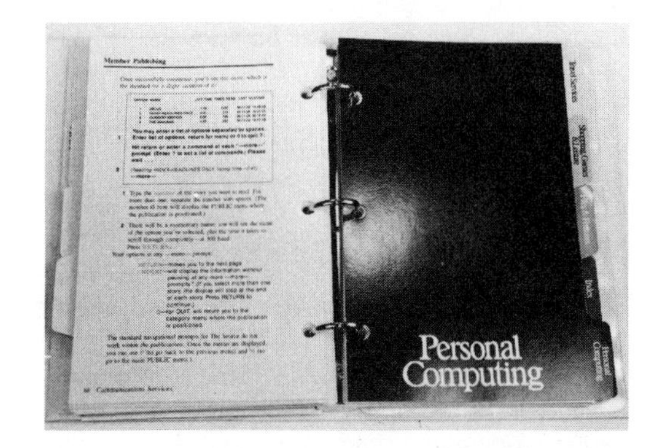

Dial-in or sign-on procedures for data services must be followed exactly. Failure to follow the rules in any user's manual will result in problems you're likely to blame on a modem malfunction.

Check the AC power to your modem first. Many "troublesome" modems have just not been plugged in.

Power cords are not the only connections to come loose. Here an RS-232C connector is not properly seated.

nected and working? If your modem is an acoustic coupler, you must be sure the rubber cups fit snugly over the two bulges of the phone handset. Exotic phone models often won't work with acoustic couplers because the two devices can't fit together properly.

Other Troubleshooting Checks

Now, let's suppose you have all these things in their correct places and that you've read (no, let's say, studied) the manuals. You sit down before the powered-up micro (and modem), boot up the communications software, dial up the service you want (using the proper phone number and access codes) — but nothing good happens. What could be wrong?

Software Checks. Are you sure you have correctly booted the communications software? Did you follow the exact instructions for the software to set the so-called "parameters"? This last question is a doozy for newcomers to the telecomputing process. We can only advise you to read your manuals carefully and do exactly what they say.

Protocol Checks. Double-check your parameter settings for baud rate, word length, parity, stop value, XON/XOFF, shift-in/shift-out and the like.

Command Set Codes. If you have the more sophisticated communications software, you will have to tell it what to do with a prescribed set of specific instructions called the command set. These commands, for example, may allow (or even require) that you "dial" the telephone by typing in the number through the keyboard of the microcomputer, using some specific "alerting" codes. Often these codes begin with a capitalized "AT" plus one or two other capitalized letters. Watch those caps. Lowercase letters don't work with some programs. Read your manuals to clarify this possibility as a source of your trouble.

Checking the Handshake with Another Computer

Call the telephone number of a known computer or data service. Listen in on the line to hear if there are computer squeals at the other end when the line answers. Keep on listening to hear if your computer and modem are responding with their own "handshaking" squeals. (This may well require you to make a response yourself, like pressing a key on the micro or a button on the modem. Read the manuals!)

Incidentally, you must wait two seconds or so after hearing the first squeal on the line before making any required response from your end. If you listen, you'll notice that the answering computer actually sends out a second and stronger squeal of a different frequency two seconds after the first. Waiting two seconds is necessary even if you're not actually listening for that second tone to appear. If you make your response before this second tone, the two computers won't handshake properly. In fact, they'll actually disconnect. Many modems have built-in speakers so you can hear the handshake. These speakers then cut out when this connection is made.

Correcting Perplexing Problems

If you feel absolutely frustrated by the problem with the new modem, the best solution is likely to be a trip to your friendliest computer store. (With luck, the one that sold you the modem and communications software!) If there is no computer store handy, look for a "hot-line" number in the modem manual. You may well find that the manufacturer's technical staff can talk you through your problem.

Try Different Software. Try a different piece of communications software. Some are easier to use than others. But again, be sure you buy software that is appropriate to your micro and your modem. Rely on a reputable computer store for advice. If you buy from a mail order concern — and there are many good ones around the country — be sure you choose one with telephone clerks who know a lot about the products they sell. Don't trust a mere order-taker who dodges your technical questions with a line like "all our customers rave about this model so you must be doing something wrong."

Try Another Modem. If nothing works, take (or send) the modem back and ask to try another brand. Be absolutely certain that the seller is a reliable source and that you have properly identified the sort of microcomputer the modem is to be used with.

PROBLEMS WITH WORKING MODEMS

What should you do to troubleshoot a modem that has been working dependably with your microcomputer and communications software?

Check the Obvious First

Be sure everything is hooked up properly, using the checks listed above for new modems. This is especially good advice if your household has children around. Busy little fingers can disconnect phone lines, power lines and cables in a twinkling — without your even knowing they've been anywhere near the microcomputer. They've also been known to damage software diskettes while computerists weren't looking. A scratched piece of communications software can behave erratically.

Check for Airborne Culprits

Household Dust. Even if your youngsters are wholly innocent, you may discover another possible cause of your problem in your household. Plain old dust in the room air! It can clog diskette drives and magnetic heads and break electrical connections.

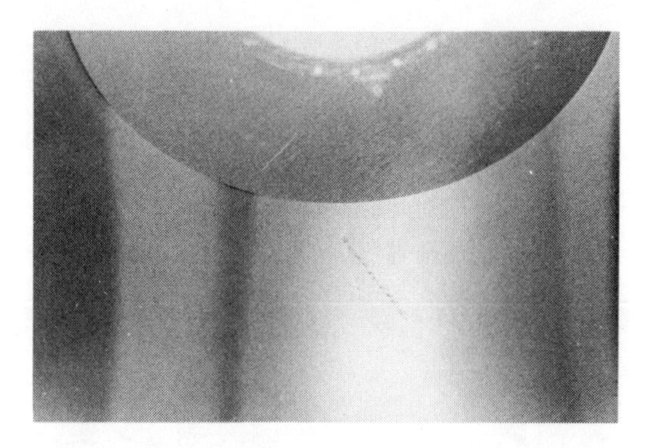

A scratched communications software disk can behave erratically.

Air Pollution. The air pollution that floats around most homes with dust can also be the culprit for a sudden, unexplained breakdown. Some experts maintain that ordinary cigarette smoke poses the same mischievous threat. Living in an urban area with high automobile traffic density can be just as bad. These airborne pollutants can leave a built-up chemical film that coats metal connections, breaking the necessary physical contacts. They can even corrode the metal in severe cases.

Clean the Cable Contacts. If you suspect dust, pollution or smoke is causing problems with cable connections, clean the metal contacts of all your wires, cables and connectors. Use a spray-type contact cleaner solution available from computer stores or electronic supply stores. Follow the general instructions on the label. Too much of a good thing can become a bad thing! Don't use this type of cleaner to get pollution gunk off disk drives or the printed circuit boards inside your computer or modem.

Preventing Circuit Destruction

Power Spikes and Voltage Surges. A serious threat to your micro and your modem is too much electricity — in the form of a massive surge or spike generated by something like a nearby lightning bolt. Lightning (and some man-made electrical problems) can damage or destroy your computer or your modem, even if they are turned off but plugged in.

The solution: proper surge and spike voltage protectors between the power supply and the

Clean metal contacts for wires, cables and connectors with a spray-type contact cleaner solution like this.

computer as well as between the power line and the modem. A separate protector is strongly advised between the phone line and the modem. Lightning can enter over phone lines as well as power lines.

An Unseen Enemy. Static electricity is another menace. The voltage generated from common, everyday static electricity around the home can destroy the microchips in a computer, modem or printer.

The solution: be sure your micro system is used only after you have "grounded yourself out" by touching a metal object or by stroking a special grounded static protector placed under or adjacent to the computer itself. Best be safe than sorry. Ground yourself before booting up.

Checking Phone Line Problems

One frequent cause of telecomputing foul-ups is the telephone circuit being used for the relay.

Check Line Quality. Listen to the sound quality of the line with your own ear at the telephone. If it seems unusually noisy or hollow-sounding, wait a few minutes and redial. Another call may put you on a new circuit or long-distance routing.

A telephone line surge protector guards against damage to your modem during electrical storms.

Try Rerouting Your Call. Telephone circuit paths can undergo radical redefinition over the span of a few minutes. If you live in Virginia, for example, and are trying to reach Ohio, your call might be routed through New York and Pennsylvania — or it might be sent by satellite to California and back by land-line all the way to Ohio. You never know — and can't request. Just play it by ear!

If All Else Fails

A last word of advice. If you have thoroughly checked your system but cannot isolate and solve the problem, sign off, take an overnight break — forget about it. Then, the next day start over from ground zero. You may well discover that everything goes together perfectly. You may never know exactly why, either. It's a part of the mystery of modern technology!

20 Telecomputing Services

For many years, technological pundits have been promising us that newspapers and letters will "soon" be delivered to our homes electronically. From time to time, we could even read about some exotic experiments here and there.

After World War II, a clumsy technique called facsimile did print out newspapers and other materials on special (expensive) machines placed in test homes and connected into slow-acting off-air receivers.

More recently, we have seen a lot of media hype about newer technologies like teletext, which uses TV carrier waves, and videotex, which uses phone lines, to transmit text and graphics onto TV screens in the home. Neither has lived up to the high-flown market promises made in their names by technology enthusiasts.

But now — thanks to inexpensive home microcomputers linked to regular telephone lines — these old promises are actually beginning to come true on a genuinely large-scale basis.

In this chapter we'll look at what these new telecomputing services offer.

The Rebirth of Electronic Newspapers. The concept caught on rapidly with the growing numbers of micro owners who were delighted to find another valuable use for their new machines. What had happened was that a working form of the electronic newspaper concept had finally been born — but without the name. These new mainframe-to-micro data services were thought of as computer resources and not journalistic delivery systems.

Almost overnight, dozens of the services were established. Many were highly specialized for groups of scientists, lawyers, doctors and librarians. But many were also designed for use by nonspecialist information consumers in their homes.

Electronic Mail

Early in the operation of these new electronic data services, the mainframe operators learned that supplying newsy information wasn't enough to keep their customers happy. They found they were also expected to make their storage-and-retrieval facilities available for text materials sent in by the micros to relay to other micros.

Electronic mail for the American home — so long envisioned — had also been born, almost by accident!

Bulletin Board Services

The appeal and popularity of these data dial-up services was so great across the country that a variation of these national efforts soon appeared in virtually every community of any size: the computer bulletin board service (BBS).

A BBS doesn't usually involve a mainframe computer. Instead, the data storage and retrieval is accomplished with an ordinary micro-

Home microcomputer workstations like this now make it possible for you to connect to data bases and services across the country and around the world.

A mainframe computer like this can store and process billions of data bits simultaneously, making data bases larger and communication faster.

computer equipped with special BBS software. Unlike the larger data services, which can deal with hundreds of calls simultaneously, an ordinary BBS can process only a single call at any one time.

More often than not, a BBS is designed for users of some particular model of micro: Commodore, Sanyo, Sinclair, IBM-PC, Apple, etc. Indeed, much of the data storage involves information about the micros themselves as well as software that dialers may download into their own machines.

Bulletin Board Access. Some BBS operations are open to all dialers. Others require a fee, often in the form of a club membership. Most require that a calling computerist register and be assigned a user number and a "secret" password so that electronic mail passed between parties can have a reasonable level of privacy. This procedure is more or less a copy of the sign-on protocol used by the big commercial data systems.

Bulletin Board Security. Confidentiality on a BBS is virtually impossible. The system operators (they call themselves *sysops*) can go poking about the stored files anytime they want to. This is much less the case in the large commercial operations where the traffic levels make it impossible for the mainframe operators to peek at much of the mail their gigantic machines process each day. Even so, if you have a message that you regard as "top secret," don't use an electronic mail delivery that is routed through a computer not solely under your control or the recipient's control. If you do, you can't be sure that prying eyes haven't exposed every word you've written.

Bulletin Board Operations. While some BBS operations are conducted full-time, others provide access only during certain hours on particular days. A few are scheduled about as erratically as "pirate" radio stations — that is, when the owner feels like it.

BBS operations that are not narrowly focused on some particular model of computer sometimes offer special sorts of information to their users. The range is great. There are BBS services that contain nothing but timely advertisements from shopping centers. Others list arts, entertainment or sports events scheduled in the area. A few run classified ads for people interested in buying or selling this or that.

Commercial data sources provide a selection from a vast information bank for the home computer user.

(Sometimes the list includes items or "services" that the law frowns on!)

Larger cities boast dozens of different BBS numbers to call. Directories of these numbers as well as the publications they spawn can often be obtained from computer stores or clubs.

Commercial Data Sources

The several large-scale commercial data services are really today's electronically delivered newspapers combined with an at-home electronic mail service. And more! Let's see what sorts of things their informational menus are likely to contain.

Access to Information on Commercial Data Sources. The main menu of a mainframe data service usually contains a listing of the main information categories you can bring into your home computer. These often include

- □ News, weather and sports
- □ Stock reports and other business-related materials
- □ News about computing
- □ Electronic mail services
- □ Electronic shopping
- □ Computer games and entertainment
- □ Travel information and reservations
- □ Computer programs to be downloaded
- □ Encyclopedia and research inquiries

Your choice will switch your screen to a second menu, which outlines the specific information subcategories available. Suppose, for example, that you have selected "News, weather and sports" on the main menu. The news menu might indicate the following subcategories:

- □ National headlines
- □ State and regional headlines
- □ Sports scores round-up
- □ Sports personality features
- □ National weather scene
- □ Local weather forecast
- □ Hollywood and Broadway news
- □ Worldwide editorial opinion

After you make your choice from this submenu, your screen will be switched to a third menu. If you had selected the subcategory "International headlines," your screen might now offer you the following choices:

- □ World happenings (English language)
- □ World happenings (Spanish language)
- □ World happenings (French language)
- □ United Nations summary
- □ Economic developments
- □ Military developments
- □ Disasters

Once you choose from this detailed sort of listing, your screen will begin to scroll the kinds of news stories you are interested in reading.

Retrieving the Information. Many people find that the letters and sentences appear on the screen too quickly to read in comfort. They prefer to command their microcomputer to "save" the incoming text on a diskette or hard disk. Then, when the mainframe transmission is over, they can play the disk back on their screen — stopping and starting it as they wish — or they can actually have the electronic file printed out on paper like a newspaper teletype.

Since you may be paying a significant sum for the long-distance connection into the mainframe, you may well decide that your local paper is, in fact, a better and cheaper way to keep up with world news. But if there is a story that catches your fancy, the computer-accessed news line can be helpful at capturing it many hours before your next newspaper is delivered.

Business people often find that the speed with which these electronic data services deliver stock market and commodity reports is extremely worthwhile. This is especially true when they're at home and away from office ticker-tapes.

Getting and Sending Electronic Mail

The electronic mail feature of most data services is easy to use if you know the electronic box number of the person to whom you wish to address a message.

Once you select the mail category on the main menu, your screen is switched to a submenu that asks you whether you want to scan (or actually read through) any mail deposited in your own electronic mail box. You also have the option of sending a message to somebody else.

Getting Your Electronic Mail. If you indicate that you wish to scan your mail, you are shown the names of the senders and the message topics. If you indicate that you are ready to read through your electronic mail, each message scrolls down your screen in turn. Many people save these messages on their computer disks for later perusal, printout or filing.

Sending Electronic Mail. Sending an electronic letter to somebody who maintains a mailbox in the same mainframe is easy.

When you select the "Send mail" option on the submenu, your screen will be switched to a sort of blank form to use. At the top is a place to type in the computer box number and/or name of the addressee. Then comes a blank into which you type the subject of the message. (This is to help your addressee scan the mail.) Finally, the screen asks you to insert the text of your message. This can be done in one of two ways, one expensive, the other cheap.

The Cost of Sending Electronic Mail. Because you are paying for telephone connections (and mainframe time) by the minute, typing a message as you sit before the keyboard of your microcomputer can become costly. Pauses, false starts and corrections take lots of time.

It is preferable to compose your message in advance, using your regular word processor. Carefully save it on disk. Then, when the screen asks you to insert the text of your electronic letter, you instruct your computer to upload the pretyped message file into the telephone line. At the end of your letter, you are ordinarily required to signal the mainframe to send it to the addressee's electronic storage box. You may also ask that the mainframe put an acknowledgment of its actual receipt in your own electronic mailbox. This is to let you know that the letter was delivered.

Many commercial data services charge your account for the time you use in sending a message. They may also bill you for the time letters from other people are stored in your mailbox. While these charges are apt to be small, it's best to clear out your box from time to time.

Another Kind of Electronic Mail

The electronic mail offered by these computer data services is quite different from the kinds offered computerists by firms like MCI Mail or by the U.S. Postal Service. Those commercially oriented mail services allow you to send mail electronically too. But, as a rule, they actually print your messages out at the geographic point of receipt and then deliver the printed version to the addressee. It's a process not unlike that used in sending a telegram.

These mail services are apt to be considerably more expensive that the less formal kinds offered by the mainframe data service organizations like the Source and CompuServe. The disadvantage in the sort of mail box services offered through these firms is that the addressee must retrieve the mail on his or her own schedule. A person who forgets to do this for several days — or weeks — becomes a "late delivery."

Electronic Messaging

There is another sort of informal, electronic messaging possible on the mainframe data service lines. One of the services imaginatively labels it "chat." It allows subscribers to write short messages back and forth as if they were tied together in a telephone conference call — or engaged in a ham radio conversation among several parties. Some people like this gossipy feature a lot; others find it trivial and boring.

Computer Teleconferencing

The same sort of arrangement can be formalized among professional people who are interested in some narrow topic. In these cases, the service is sometimes labeled "computer teleconferencing." It permits people to communicate complex ideas in writing without the long delays of regular correspondence. Many times computer teleconferencing links go through a network of mainframe computers, allowing you to access the system anywhere in the country by placing a local call to a member computer — usually at the campus of a large university.

These computer teleconferences are ongoing brainstorming sessions where each participant can add information to what the others have left. They are also used by business for more routine communications, like leaving orders from a sales staff in the field, or relaying company memos to people on the road.

Travel Information and Reservations

The travel information and reservation services available from the mainframe data services provide an unusual convenience for people who do a lot of traveling by plane or train.

If you select this service category at the main menu, your screen will be switched to a submenu that might contain the following choices:

- □ Airline schedules
- □ Train schedules
- □ Travel reservations
- □ Hotel information
- □ Hotel reservations
- □ Restaurant information
- □ Restaurant reservations

The process usually involves your looking up schedules and costs, then booking the travel arrangements you need. As a precaution, you should save the reservation confirmations that will appear on the screen at the end of the booking process. Print them out and take them with you. They're your best evidence of commitment by the various airlines, hotels and restaurants you have booked.

Electronic Shopping

Electronic shopping services work in much the same way as travel services. You look through catalog listings of all sorts of products: computers, cameras, hi-fi equipment, sports gear, etc. Each such listing contains a purchase price, usually attractively discounted over the cost of the same item in a retail store.

When you've made up your mind, you type the required order information, including your credit card number and mailing address. Save and print out the order confirmation as a record should problems arise later.

Electronic Banking

If your bank allows computer transactions of your accounts, you dial up the bank's own mainframe and follow the bank's specific instructions for transferring funds, paying bills and the like. As a rule, you won't work through a data service computer operated by a third party. This is because you and the bank want your financial transactions to be completely private and confidential.

Even computer games are available on many mainframe data services. Some services let you download these games to your computer (for a fee) to use without running up your bill.

Computer Games

The computer games available through the mainframe data services range from simple word games to arcade types. Many are designed to be played between the user and the mainframe. They consume lots of (chargable) time. A few are designed to be downloaded and saved in your own computer. As a rule, these are not likely to be as appealing as the games you buy as software in a store. But they are apt to be a lot cheaper.

Educational Data

From time to time, you may discover that educational courseware is actually available from a source like a mainframe data service. More likely, these highly specialized materials can only be obtained from mainframes operated by large universities.

The materials are ordinarily designed around the principles of programmed instruction. You read a short explanation, followed by a pertinent question. You answer. If you are correct, you move on to the next "frame." If your answer is incorrect, you are routed to a "remedial branch," which helps you learn the material you missed.

These educational materials range from math drills for youngsters in the elementary grades all the way to college courses in economics and calculus. One of the more popular educational services is the SAT preparation lessons (for high school students preparing to take the Scholastic Aptitude Test for college entrance). Several colleges now offer through your computer most of the courses you need to earn a degree. With these courses, you are not only given information and drill in the subject but have a computer teleconferencing link with your instructor.

If you're interested in any of these educational uses of computer data sources, contact the continuing education department at your state university or community college to see if any really good materials are available in your area.

Electronic Research

Another, more practical sort of educational service often available from mainframe data services could be called a "research" function. You type several words that indicate your interest in learning more about some particular topic, like aviation, computers, flowering bulbs or Austrian history.

In a relatively short while, your screen shows text materials, or bibliographies, drawn from one or more encyclopedia entries on the topic requested. Downloading the ensuing data in your own microcomputer system will allow you to use them later in academic themes or research papers.

These lookup services are proving helpful to families with children in school. Using them is often much easier than driving to the local library, particularly late at night when the assignment has "suddenly" been remembered.

Added Cost for Added Services. Please note one warning about some of these data services. Some — like the travel, shopping and research examples — may cost you an extra fee. Be sure you investigate before using them and then discovering that your monthly bill reaches orbital heights!

The history of these new electronic services for the home is just beginning. We can all expect to see many exciting — and inexpensive — developments in the years ahead.

21 Picture Telephones

During the 1964–1965 New York World's Fair, Bell Laboratories introduced an experimental telephone that transmitted not only voice but full motion pictures over telephone lines. Since then, the concept of picture telephones has been vacillating between comic book fantasy and real-world practicality, looking for a place and a way to stabilize into commercial application and acceptance.

The general public viewed the new Bell invention with controlled enthusiasm. There were those whose vision about the new technology stopped abruptly at the thought of answering the picture phone in the middle of their morning shower. On the other hand, families that grew more separated as society became more mobile saw the picture phone as a way to draw the family unit closer together again. Grandma could now see the new baby, not just listen to gurgles and burps. Had Bell Labs' new invention caught on then, the whole concept of "reach out and touch someone" would now have a quite different meaning.

The picture phone was introduced by Bell Laboratories at the 1964–1965 World's Fair. Use of this technology is becoming more common in business applications and recent developments make home use of the picture phone much closer to practical reality.

The last two decades have seen very little progress in the development of the picture telephone for the general public. In these same two decades, however, giant strides have been taken in this technology for business applications in a whole new concept called *video teleconferencing*. Because of the increasing sophistication to meet demands from the corporate sector, the reality of having a picture telephone in your kitchen is now just around the corner—maybe even closer than that.

Why has the technology taken so long to emerge into the market place? The answer is twofold: cost and speed. To transmit pictures fast enough to be perceived as full-motion video, special high-cost telephone trunk lines had to be used and dedicated only to this service. The $1000 to $1500 per hour charge for these special lines took the picture telephone way out of reach of the ordinary household.

To understand the engineering dilemma, you have to know a little about how picture telephone technology works.

Electronic Pictures

A video picture is made by sending a very thin stream of electrons to sample light and dark patterns of an image that is focused by a lens on the face of the camera tube. The electron beam is concentrated into a small dot and starts at the top left-hand corner, moving across the screen in a more or less horizontal line. When the electron beam finishes the first row, it is turned off and aimed again at the left side of the picture to start another scan line below the one it just completed. After 262 lines it is at the bottom of the camera tube. Total elapsed time to get there: $\frac{1}{60}$ of a second. Somewhere in the middle of line 263, the beam is turned off and snapped back to the top of the tube to scan the spaces between the lines it just did. In just $\frac{1}{30}$ of a second the electron beam scans 525 lines down the image area.

At about 200–400 regular intervals on each scanned line (the number varies with the sophistication of the system), the electron beam detects the location and intensity of light and converts these to voltage differences. Light areas produce a higher voltage output than darker areas of the image. These voltage changes are sent to the associated camera equipment circuits for processing and transmis-

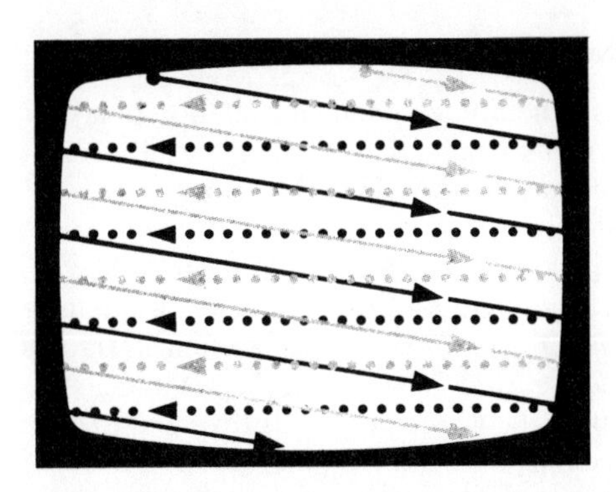

An electron beam scans the face of a TV tube with 525 left-to-right lines running from top to bottom of the screen. The 525-line frame is composed of two interlaced fields of 262½ lines each. The electron gun begins at the top left corner of the screen, zig-zags to create the first field and then races to the top center to trace the second field between the lines of the first. The number of these 525 lines seen on a TV tube defines the *horizontal resolution* of the tube.

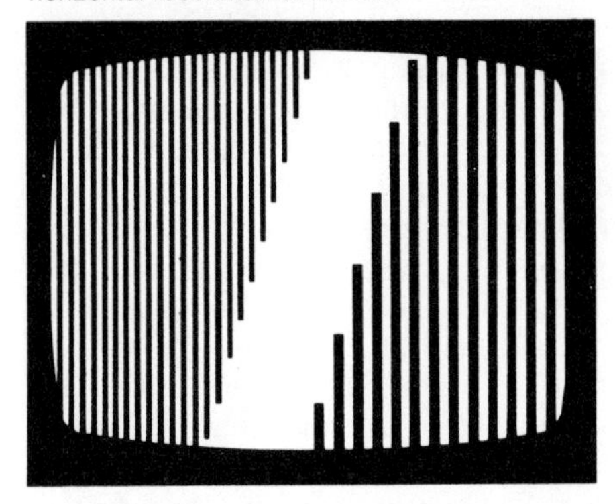

The intensity of light focused on the pickup tube is sampled by the electron beam during each of the 525 lines it scans. High-quality, high-definition pickup (and reception) tubes will sample the light intensity 300–400 times in the very brief time it takes to scan one line. Most home television sets will do this only about 200–250 times, resulting in a lower definition picture. This is the tube's *vertical resolution* because it defines how many vertical lines of dots are displayed.

sion over cable to your VCR or over the airwaves to your TV set.

Each of the dots sampled in the pickup tube represents a separate voltage cycle. Each frame scanned by the electron beam is made up of hundreds of thousands of these sampled dots. With 300 dots sampled on each line in a pickup tube, and 525 horizontal lines per frame, there are 157,500 dots per frame. Each second, 30 of these frames are sent out. This means the transmission medium has to handle about 4,725,000 voltage changes each second.

The transmitting medium must be able to handle that much information that fast. Ordinary telephone lines can't. A normal voice channel on a telephone line can handle only about 3000 of these voltage cycle changes a second. To transmit a single TV transmission, over 1400 telephone lines would have to be used. Even if that were technically possible (and it is not), it would not be practical. Outside the TV networks (which used land lines to get their programs across the country before satellite transmission was available), there was not much enthusiasm in the home or corporate sector for investing in special leased telephone lines to transmit video at a rate of $1000 to $1500 per hour for full-motion pictures over telephone wires.

A Still Picture Compromise

Engineers could send the same amount of information needed to make up each television picture, if they sent them out at a slower rate that ordinary phone lines can handle. The result: still pictures instead of full-motion video.

It soon appeared that the public was willing to accept this compromise — at least business and corporations that needed to communicate both voice and images to scattered segments of their company were. They also accepted the fact that the pictures would not be in full color. Color TV requires much more information be sent in the same time interval to get an image of acceptable quality.

Voice Teleconferencing

In the corporate world, sensitive to growing expenses and shrinking productive time, teleconferencing quickly became a way of life. The manufacturing division in Dallas could connect immediately by telephone with corporate headquarters in New York and sales offices in Chicago and Seattle. Business was conducted and decisions made without anyone having to pack a suitcase or get on a plane. The whole decision-making process could be done within 15 minutes and you'd be back at your desk working on your new assignment.

Even education was quick to pick up on this concept, sending classroom sessions to "homebound" students in two-way communication with the teacher and classmates by speakerphone.

But voice teleconferencing did not meet the need of companies that wanted to share architectural or engineering drawings or get an expert's advice on a broken component — especially if the expert was in Paris and the part was in New York.

Videoconferencing

Now it is possible not only for corporations but for education and the professions to share visual information with almost the same economics as voice conferencing. A doctor in Virginia can get immediate assistance from a specialist in Minnesota about an X ray or EKG. A community college with a widely scattered student body can set up videoconferencing centers in any business or community location, giving the campus hundreds of "classrooms" within walking distance for the commuting student. And for the corporate world, new product lines can be introduced, production processes and new ad campaigns shown to all branch segments quickly and cheaply without anyone having to travel to a central site.

For the home owner, the picture telephone in the kitchen is much closer to reality because of these forays into innovation and experimentation by the corporate and educational sectors of society.

Video teleconferencing is being used in colleges, where more classes can be made available to a widely scattered student body or where specialized faculty can be shared among campuses and other colleges.

Picture Telephone Technology of Today

A typical picture telephone installation today is a marvelous application of the state-of-the art technologies of television and computers.

Basically, the system consists of a TV camera to pick up the video image, a processor to convert the picture to digital values, a modem to transmit a more economical analog version of these digitized images, and the receiving station where the whole process is reversed to display the image on the TV tube.

In most systems today, especially those for the home market, the picture telephones display still pictures in black and white, not color. The amount of information that must be transmitted is much less with the tonal shades of gray than with the full palette of hues and tones for each of the three primary colors of light. Even monochrome pictures are limited to about 16 shades of gray. Still pictures are more economical to use than full-motion video. While some systems do offer full-motion, full-color video, the initial cost of the equipment and increased telephone line charges put them way out of reach of the average family.

For the average consumer, hookup is simple. A connection is made to the telephone lines through a standard RJ-11 modular plug. The power cord is plugged in and the unit turned on. A call is then placed to someone with a similar unit (an obvious necessity, of course, is at least two units). You talk to your party over a standard telephone, or over a microphone plugged into or built into the picture telephone unit. You then press one or two buttons — usually on the electronic-tone dial pad. This takes a snapshot or "freezes" a picture picked up by the TV camera. (This is why the technology is often called "freeze-frame television.") The picture is digitized so the built-in computer can understand tonal values of various dots that make up each image. It is then sent through a special computer modem along ordinary telephone lines. At the receiving end, the modem there converts the picture back to digital data, where it is processed into the analog tones and sent to a high-resolution TV screen for display.

Because the information is sent out bit by bit as modems do, it takes a while to "write" the picture on the screen. At the receiving site, you see it "wipe" on the screen from top to bottom as each bit of the picture goes through the translation and is converted to a screen image. It can take anywhere from 1 to 20 seconds to complete the transmission and display of a picture, depending on the detail in the picture and resolution or quality desired.

As soon as the transmission of one picture is complete, another can be sent immediately — from either end of the picture telephone connection.

Basic Unit for the Home

The newest version designed for the home market is the Luma Phone™ by Mitsubishi. It uses a 3-in. display screen with a fixed camera built into the unit and aimed at the person operating it. The screen can display images from both the sending and receiving site.

Images can be transmitted and received by the unit in 1–5 seconds. The resolution of the picture is not good enough to display printed text or drawings in a readable form. When displayed on the unit's built-in 3-in. display tube, however, the black-and-white image of a person's face is of acceptable quality to all but the pure videophile.

The unit plugs into regular telephone lines with an RJ-11 modular jack. The telephone can, of course, be used by itself when calling someone without a similar Luma Phone unit.

The Luma Phone, at about $1500, is almost priced low enough to begin competing with voice-only telephones.

Photophones for Videoconferencing

Image Data, a Texas-based corporation, has more than 400 similar, but more sophisticated systems in place — mostly in business and education — with a price tag of about $8500 per unit.

Image Data's Photophone™, with its 9-in. screen and electronics that give much higher resolution, can transmit readable text and detailed drawings. This increased picture quality is a necessity in business and education applications, where you might need to send an engineering drawing or a picture of an X ray or the insides of some faulty component to some remote location in the country for instant analysis or use.

Luma Phone controls are simple and the unit fits comfortably on a desk. At the press of a button, a TV picture of you is "frozen" and then transmitted by ordinary phone lines to the receiving Luma Phone. The fixed camera is aimed by moving the whole unit.

The Luma Phone™ has a 3-in. display screen that shows what your camera is sending and the still picture being received from the person you're calling. This is one of the first picture telephones priced low enough to compete for business, and even some home use.

The Photophone™ by Image Data has a 9-in. screen and electronics that give it high enough resolution to read text and detailed drawings.

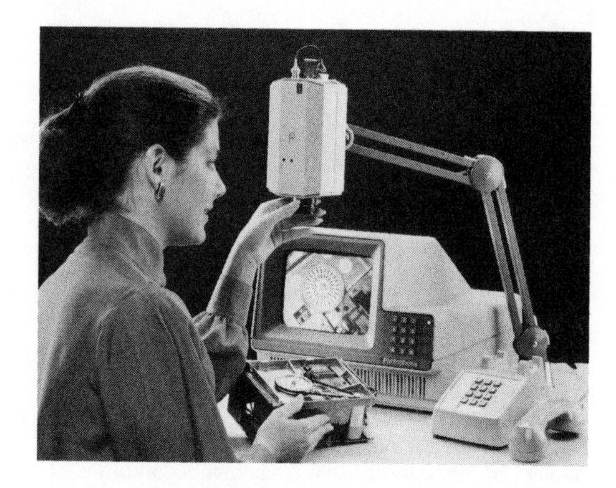

In business and manufacturing applications, problems can be solved almost instantaneously by sending a picture of a troublesome part to a troubleshooter.

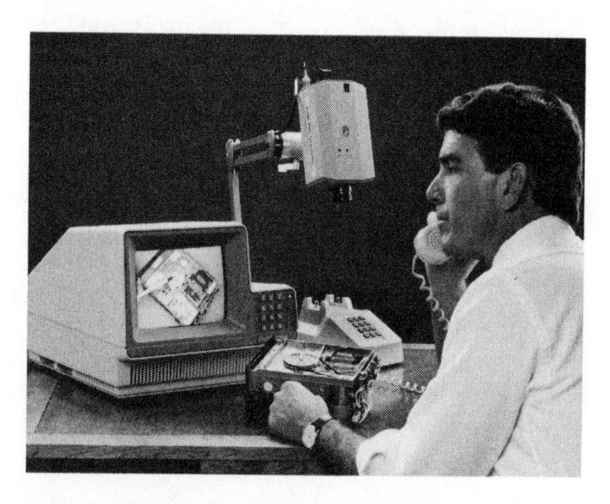

Engineering experts in another plant location or another city can quickly analyze the Photophone picture of the part and give corrective procedures on the spot.

The Photophone has a camera mounted on a movable arm that can be turned to aim at different people, or turned vertically to show written messages, drawings or close-ups of objects resting on a table.

The Photophone automatically adjusts its transmission rate to the quality of the phone lines you are using. If it can't successfully send the image at 9600 image bits per second, it senses the error before displaying the image and drops to a lower speed of 7200 or 4800. If the phone lines are too bad for Photophone to send an acceptable image, it tells you to hang up and try again, perhaps to get a different circuit or routing for the call. The time it takes to write a full-screen picture ranges from 7 to 20 seconds, depending on the resolution and how fast the phone line quality lets it be sent out.

Directions for using the keypad to control Photophone operations are given on the lower corner of the screen. You press only a couple keys to send a picture.

There isn't much need for an instruction book with the Photophone. All operating directions can be seen in the lower right corner of the screen in the 12-square matrix of the electronic tone pad used to control operations.

After placing a call to a receiving station, you aim the camera by moving it on its extendible arm. You can transmit images immediately, or store them temporarily in the computer's memory or permanently on a 3½-in. floppy disk.

Standard video inputs let you use the Photophone with a VCR or videodisc to capture and send images. A standard video output lets you feed a video tape recorder, a larger screen monitor or a video distribution system throughout a plant, office or classroom building (or home).

The internal disk drive operates like a slide projector. Before placing your call, you record the pictures and then arrange them electronically in the order you want to send them. When you place the call, changing images is as easy as operating a remote-control slide projector — although changes between pictures are slower by the amount of time it takes to send the picture and write it on the screen.

Voice and video transmission are done over the same telephone line. Voice communication is interrupted briefly in the time it takes to use the line to transmit a picture.

The size and portability of the Photophone unit has given it much greater use in business and education teleconferencing applications. Designed to occupy about the same "footprint" on a desk as a personal computer, the unit is light enough (32 pounds) to be carried to someone's desk or workstation, eliminating the need in most business and education applications to

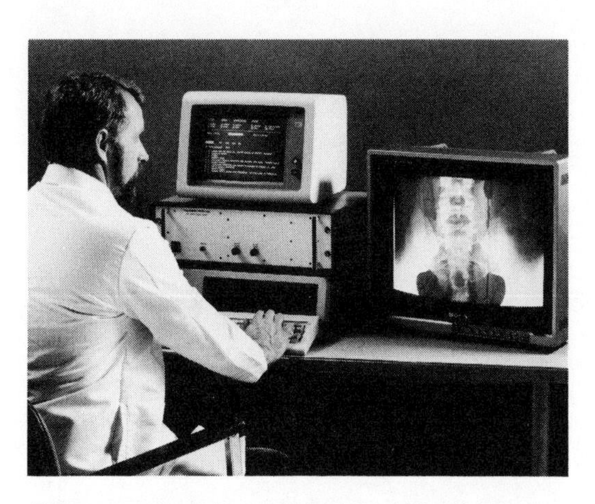

This system by Colorado Video is typical of the more sophisticated systems used by business in freeze-frame video conferencing. It has high resolution for detailed pictures but depends on you having the pickup cameras and other originating and receiving equipment.

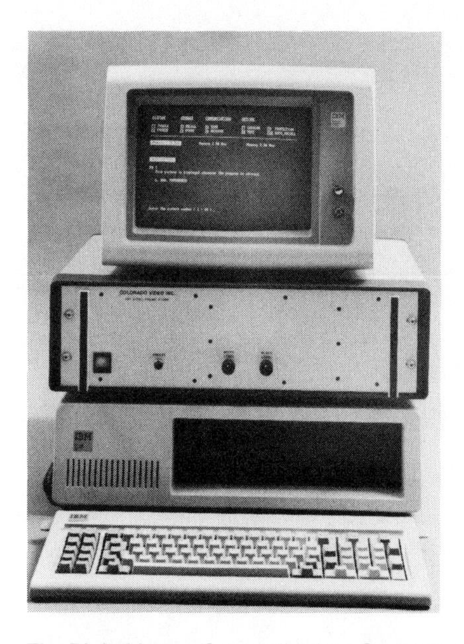

The Digital Image Communications System™ can store and transmit full color images. The components shown here are used with an IBM-PC personal computer and require a special high-band telephone line to transmit the picture information.

set up large and expensive teleconferencing centers or dedicated rooms.

Larger Teleconferencing Systems

An early pioneer in the field of freeze-frame video conferences is Colorado Video in Boulder. Its system contains only the digitizing, transmitting and receiving units, which work similarly to those described earlier. It depends on your installation to have its own camera and other equipment to capture the images and send them to the unit for processing through standard video connections. The "freeze-frame" unit then converts the images for transmission over regular phone lines. The company using these units usually has a special room dedicated to teleconferencing use where the rack-mounted equipment and associated TV gear can be permanently installed.

The Digital Image Communications System 950™ by Colorado Video does store and transmit full color images. The system is used with an IBM-PC personal computer. The 10-MB hard drive will store four full-frame color images for later transmission. In contrast, the same system can store 160 low-resolution or 40 high-resolution black-and-white pictures in the same disk space it took to store four color pictures. This shows dramatically how much more information is needed to be stored and transmitted for full color. It also explains, in part, why picture telephone technology remains largely monochrome.

With this system, you need two telephone lines for the conferencing hookup: one to transmit voice, the other to transmit the pictures. The company also markets an accessory instrument designed to transmit still video images during the vertical blanking interval of a television broadcast station. (The vertical blanking interval is the very brief time the transmission of information for each scanned video frame stops and the transmission of a new frame of picture information begins. It happens so fast, you never notice that the screen is blank during this brief interval.) Using only part of this vertical blanking interval time to transmit information, a color image can be sent in about 8 seconds.

Implications for the Home

These advances in the technology are not of much value to you as a do-it-yourselfer who does not operate a broadcast TV station in your home or does not have the immediate need for videoconferencing in your business. It does, however, point out advances in the picture telephone technology in operational use now at the higher level industrial, commercial and educational applications. Eventually these will distill into the cheaper, mass-produced units used in your home.

Index